L. I. H. E.
THE MARKLAND LIBRARY
STAND PARK RD., LIVERPOOL, L16 9JD

© D. H. Cushing 1977

First published 1977
by Edward Arnold (Publishers) Ltd
25 Hill Street, London W1X 8LL

Boards edition ISBN: 0 7131 2673 6
Paper edition ISBN: 0 7131 2674 4

All Rights Reserved. No part of this publication
may be reproduced, stored in a retrieval system,
or transmitted, in any form or by any means electronic,
mechanical, photocopying, recording or otherwise, without
the prior permission of Edward Arnold (Publishers) Limited.

Printed in Great Britain by
The Camelot Press Ltd, Southampton

General Preface to the Series

It is no longer possible for one textbook to cover the whole field of Biology and to remain sufficiently up to date. At the same time teachers and students at school, college or university need to keep abreast of recent trends and know where significant developments are taking place.

To meet the need for this progressive approach the Institute of Biology has for some years sponsored this series of booklets dealing with subjects specially selected by a panel of editors. The enthusiastic acceptance of the series by teachers and students at school, college and university shows the usefulness of the books in providing a clear and up to date coverage of topics, particularly in areas of research and changing views.

Among features of the series are the attention given to methods, the inclusion of a selected list of books for further reading and, wherever possible, suggestions for practical work.

Readers' comments will be welcomed by the authors or the Education Officer of the Institute.

1977

The Institute of Biology,
41 Queen's Gate,
London, SW7 5HU

Preface

Some commercial fisheries have been studied scientifically for half a century or more. During that period exploitation has increased steadily and nations have co-operated in certain parts of the world in controlling the effort exerted upon the fish stocks. In the last decade or so fishermen in large and complex vessels have steamed further and further from their home ports to find rich stocks of fish. Foreign fleets have been seen off the shores of tropical and subtropical countries and as a result more nations throughout the world declare economic zones out to 200 miles in line with the theme now developing in the UN Law of the Sea Conference. However, during the period since the Second World War much of fisheries science has developed in an international forum.

Some of the scientific problems have been solved, but their solution not unexpectedly reveals further ones. In general, satisfactory solutions to fisheries problems have not emerged until perhaps ten years of data have been collected. Today, because of the pressure of fishing, difficult problems emerge in a much shorter time span, as for example in the Peruvian anchoveta fishery which collapsed from unknown causes after about a decade of heavy fishing. This book seeks to state the present problems facing fisheries biologists.

Lowestoft, 1977   D. H. C.

Contents

	General Preface to the Series	iii
	Preface	iii
1	**The Environment of the Abundant Fish Stocks**	1

1.1 Introduction 1.2 The physical environment 1.3 The structure of life in the sea 1.4 The production of life in the sea

2 **The Biology of Fishes** 17

2.1 Introduction 2.2 Growth and mortality during the life cycle 2.3 The circuit of migration and reproductive isolation 2.4 The generation of the recruiting year classes

3 **The Population Dynamics of Fishes** 28

3.1 Introduction 3.2 Estimation of fishing mortality 3.3 The Graham-Schaefer model 3.4 The yield per recruit model 3.5 Recruitment and parent stock 3.6 Conclusion

4 **The International Management of Fisheries** 38

4.1 The early history of international management 4.2 The International Council during the thirties 4.3 The development of conservation during the fifties and sixties 4.4 Management today

5 **Science and the Fisheries Today** 47

5.1 The structure of currents in the sea 5.2 Catch statistics and population dynamics 5.3 Independent estimates of stocks by acoustic methods 5.4 Long term trends in the climate 5.5 Fish farming 5.6 Pollution 5.7 Conclusion

References 59

1 The Environment of the Abundant Fish Stocks

1.1 Introduction

The world catch of fish from the sea each year amounts to nearly sixty million tons. The Russian, Japanese and Peruvian fishermen catch most, followed by those from Norway, the United States, Spain, Thailand, the United Kingdom and Chile. In both Russia and Japan the deliberate development was started some decades ago to ensure adequate supplies of protein and today their fleets search for trawled fish and tuna in all oceans. Off Peru a purse seine fishery was started for anchoveta in the late fifties and early sixties and until 1971 it yielded about ten million tons each year, nearly all of which was converted to fish meal; since then catches have been much reduced. During the sixties a trawl fishery was developed in the Gulf of Thailand which now yields about a million and a half tons each year. Norwegian fishermen catch a similar amount of capelin (a small smelt-like fish) in the Barents Sea; a few years ago the same fishermen took 0.7 million tons of mackerel in the North Sea. Such are the major developments in the last decade. A recent estimate (GULLAND, 1971) suggests that not more than about one hundred million tons could be caught annually from the world ocean with conventional gear and that this limit might be reached within the next decade or so.

There are four traditional methods of catching fish: by hook and line, by drift net or gill net, by purse seine and by trawl (Fig. 1-1). In days gone by, cod were caught in the North Atlantic by about two miles of line with baited hooks every fathom or so; the great lines, as they were called, were shot and left on the sea bed for a period of a day or so before they were hauled. Today, although the method is very labour intensive, it survives in many inshore fisheries and in the Pacific halibut fishery off British Columbia and Alaska. For the highly valued tuna the Japanese use unbaited hooks on pelagic long lines of up to 80 km in length from each boat and they fish in all parts of the tropical and subtropical ocean. The drift net or gill net is a long curtain suspended from the surface by floats and it drifts or drives with the tide and fish swim into it and are caught by their gill covers. (A set net is similar to a drift net but is anchored to the sea bed.) Before the Second World War most of the North Atlantic herring were caught by drift nets, but these have been replaced by more efficient trawls and purse seines. However, gill nets of up to 80 km in length are still used by Japanese fishermen to catch salmon from Asian rivers in the ocean south of the Aleutian Islands.

The purse seine is a ring of net that is shot to surround a shoal of fish in

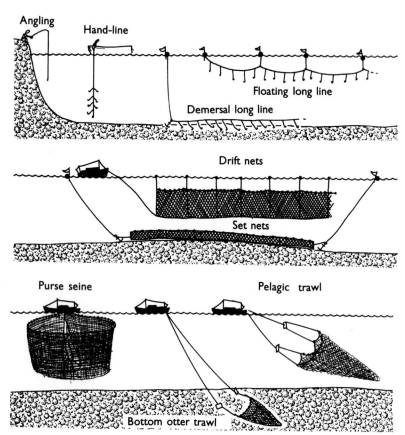

Fig. 1-1 Some of the gears used in the fisheries; purse seines, pelagic trawls, drift nets and floating long lines are used in the midwater, whereas bottom otter trawls, set nets and demersal long lines are used on the sea bed.

midwater and is closed underneath with a 'pursing ring'; it is very efficient especially when used with sonar to locate the shoals. The purse seine is used in the Peruvian anchoveta fishery, the Norwegian capelin fishery and in the southwest African sardine fishery, amongst others. The trawl is a conical bag dragged along the bottom; its mouth is spread by otter boards at the ends of the two towing warps.

During the first half of the present century most of the fish caught was sun dried (like the cod exported from Norway to West Africa) or was preserved on ice, for example, for the fish and chip trade. There were limited markets close to ports for fresh fish from inshore vessels. Although such outlets remain, two major processes were introduced during the fifties. The new trawlers built since then fish from the stern rather than over the side as did the older ones and the catch is deep-frozen at sea and is processed, for example as fish fingers. The second process involved the production of meal; because vegetable protein lacks the essential amino acids, lysine and methionine, the broiler chicken industry was built on animal protein, the cheapest form of which was, and is, fish meal. Much of the fish that goes into meal is caught by purse seiners, which ensures continuous and efficient throughput to the factories.

The major fisheries in the world are found on the continental shelves, particularly where they are broad, for example, in the North Sea, the Barents Sea, the Bering Sea, the East China Sea and the Grand Banks. On the eastern boundaries of the subtropical oceans, there are upwelling areas where cool nutrient-rich water rises from 100 to 200 m; purse seine fisheries for sardine-like fishes and trawl fisheries for hake occur off northwest Africa, southwest Africa, Peru and potentially off California.

A fishery is found on one of three characteristic grounds, where fish feed, where they spawn or where they winter. The trawl fishery for cod on the Svalbard Shelf (south of Spitzbergen) in summer, is a feeding fishery where the cod gather to feed on krill (a small shrimp-like animal) or capelin. A line fishery for cod takes place in early spring on their spawning ground in the Vestfjord in northern Norway. There is an overwintering fishery for mackerel by purse seiners in the deep water off the southwest coast of Norway. Fisheries occur at regular seasons and at recognized positions every year in what appears to us a featureless ocean.

1.2 The physical environment

The sea is salty (30–35‰) and usually rather cool. In high latitudes and at great depths it is cold ($-1°$ to $3°C$), but the surface layers are heated by the sun; in summer, tropical surface waters may reach $30°C$ in places, temperate waters $15°–20°C$, yet in high latitudes the summer surface temperature may be as low as $5–7°C$. In depth, there is a boundary to the heated layer, called the *thermocline*, with cool water below. During the autumn in temperate waters, storms stir the water column sufficiently to destroy the thermocline and the column becomes isothermal; in the following summer, as the upper layers are heated, the discontinuity, or thermocline, becomes re-established. The depth of the seasonal thermocline depends on the heat received, the depth of water and the degree of vertical mixture; if the tides are swift no thermocline is established in summer, as for example, in the eastern English Channel and the southern North Sea.

Although most of the mineral content of the sea is salt, most elements are present and some ions in particular form the nutrient quality of sea water, for example, phosphate, nitrate and silicate. Oxygen is present in sufficient quantity to support life nearly everywhere and indeed near the surface the water is frequently supersaturated. But in the deep ocean between 400 m and 800 m there is an oxygen-poor layer. Carbon dioxide enters the water from the air, is dissolved and is buffered by the bicarbonate ions; the average concentration of carbon dioxide is about 90 mg/l. It provides the principal fuel for photosynthesis, together with other nutrients, which supports all life in the sunlit water of the sea.

Light is absorbed and scattered in sea water and the depth of greatest penetration in the clear water of the deep ocean is about 800 m; below that depth is a zone of bioluminescence lit by light-producing organs or photophores on the cephalopods, fish and crustacea that live there. Sea water is a filter and the white light at the surface is reduced to a greenish blue with depth. Nearly all photosynthesis occurs in the *euphotic layer* down to a point in depth at which the surface intensity is reduced to 1%; 10–25 m in the North Sea, 60–100 m in the Sargasso Sea. The algae tend to live in the lower half of the euphotic layer and light used for photosynthesis is greenish blue.

The visual range of an animal in the sea is limited by the scattering of light and indeed a prey target is detected by a predator by its contrast against the scattered light. Both prey and predator must seek to minimize the contrast with blue or grey colours or with silvery scales that reflect light away from the horizontal. In the clear water of the deep ocean, the maximum visual range may be as much as 80 m, but in more turbid waters like the southern North Sea, it may be as little as 3 m. The eyes of fishes are adapted to the colours in their somewhat contrastless environment and their visual pigments are most sensitive to blue-green light.

Like the atmosphere, the sea is always in motion and for the same reasons. Both systems respond to the earth's rotation and the stresses of wind across the ocean play a part, but only a part, in maintaining the major surface current systems. Another and major component in the generation of the currents is the differences in heat budget in the surface layers across the latitudes. The largest circulatory systems are the anticyclones between the equator and about 40° of latitude (Fig. 1–2). An anticyclone in the northern hemisphere is a clockwise circulation just as in the atmosphere, and in the same way it is anticlockwise in the southern hemisphere. The major anticyclones are formed of a westbound equatorial current, one at the western boundary which is swift and sharply defined (for example, the Gulf Stream in the North Atlantic and the Kuroshio in the North Pacific) and which spreads eastwards along latitude 40°, and a slow diffuse eastern boundary current (for example, the Canary current or the California current). Between the two equatorial currents, north and south, runs the eastbound equatorial counter-current at about 8°N. In the Pacific, the south equatorial current lies just north of the equator. At the equator itself, at a depth of about 180 m, there is a fast but narrow undercurrent that flows eastwards to the eastern ocean.

Poleward of latitude 40°N lie centres of anticlockwise (cyclonic)

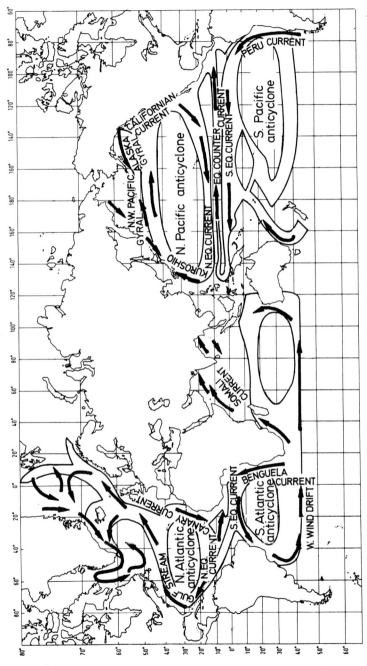

Fig. 1-2 Current systems in the world ocean. (From *Encyclopaedia Britannica* World Atlas, with permission from *Encyclopaedia Britannica*, Inc., Chicago.)

circulation. In the Pacific there are two, the Alaska gyral and the North West Pacific gyral, and in the Atlantic the circulation in the Norwegian Sea is also cyclonic. In the southern hemisphere, because there are no land masses between Antarctica and South America and Southern Africa, no cyclonic circulations are established and they are replaced by the West Wind Drift; the whole mass of surface water moves in an easterly direction under the westerly winds.

In the eastern boundary currents of the anticyclones, extensive areas of upwelling occur, off California and Oregon in the California current, off Peru and northern Chile in the Peru current, off northwest Africa between Dakar and Casablanca in the Canary current and off southern Africa between the Cape of Good Hope and the border of Angola in the Benguela current. In all these regions a wind blows towards the equator along the shore or at an angle to it. In the northern hemisphere, a mass of water in motion is turned to the right under the influence of the earth's rotation and in the southern hemisphere it is turned to the left; this effect is called Corioli's force. Under the influence of this force and the stress of the longshore wind, the water is moved offshore and it is replaced at the shoreline with water from below, which is cool and rich in nutrients. This water is dragged up from as much as 200 m. The region of upwelling may extend over a considerable stretch of coastline and it shifts slightly poleward as spring turns into summer. There is a dynamic boundary about 50–100 km away from the coast where the water moved offshore meets the main eastern boundary current. Because of the high transfer rates of nutrients, the regions of upwelling are regions of high production and high fishing activity.

If we examine the physical environment through the senses of fishes the sea must be as structureless and featureless as it appears on its restless surface. As already noted, the visual field is an empty blue grey with some differences in contrast as prey or predator crosses it. Little is known of the sound field save that there must be a continuous low frequency noise from the surface, especially in coastal waters. Fish have a considerable capacity for taste and smell and the water provides a convenient carrier for a complex array of ions. The tides and currents probably carry the animals willy-nilly unless they have the external referents obviously available to fish that swim up rivers. However, fish are aware of the water flowing across their bodies and can detect the movement of a neighbour in a shoal. Despite the limited sensory information, fish find their way from place to place in the sea with considerable regularity, as will be shown below.

1.3 The structure of life in the sea

The plants that live in the sea are among nature's smallest and, excluding the seaweeds, the largest is not more than 1 mm across (Fig.

1–3). There are small naked green flagellates, spherical or elliptical in shape that range in size from 1–10 μm. Coccolithophores are small oceanic flagellates with calcareous plates about them. Diatoms are siliceous pill boxes, sometimes in chains, sometimes solitary and some have long tubular extensions that help suspend them in the sea; they range in size from 1–100 μm across, with a few larger individuals here and there. Dinoflagellates are about the same size and they are motile and may be naked or armoured with cellulose. All such algae bear chlorophyll and perhaps masking pigments, and they live in the euphotic layer with light, carbon dioxide and nutrient salts for photosynthesis and growth. They sink slowly in still water (0.6 m/d on average) but are usually sustained by turbulent mixture in the water.

The small animals that feed on the plants are predominantly copepods, small crustaceans 0.3–3 mm in length. They capture the algae and filter them from the water and those that feed on large diatoms have siliceous 'teeth'. As might be expected there is a considerable size range of such animals, from radiolarians and tintinnids (protozoans) to the larvae of copepods (nauplii), of molluscs (veligers) and of salps such as *Oikopleura*. There are planktonic carnivores that feed on copepods, for example, arrow worms, *Tomopteris* and ctenophores. However, fishes are the prime carnivores in the sea. Herrings and sardines cruise slowly in shoals selecting and filtering copepods from the water; each may eat as many as a thousand in a day, so a fair proportion of their time is spent in feeding. In contrast, a predatory cod, swimming alone, attacks a whiting at high speed and may not make more than a dozen such attacks in a day. Cod may also feed on capelin or on krill and in the deep ocean the tuna-like fishes feed on myctophids, little fish with photophores. The top predators in the sea are the large sharks, dolphins and toothed whales. Even larger are the huge non-shoaling filterers that take plankton animals by the ton each day, the basking shark, the manta ray and the baleen whales.

The productive engine in the sea is driven by the photosynthesis of algae in the euphotic layer and it is supported by the uptake of carbon dioxide and nutrients there. At the depth of 1% penetration of the surface light intensity, the rate of photosynthesis equals the rate of respiration in the algae and new production of organic material can only occur above this *compensation depth*, i.e. where the rate of photosynthesis exceeds the rate of respiration. Photosynthesis increases with light intensity, exponentially with decreasing depth, but close to the surface it may become inhibited; the distribution of production in depth, measured in units of chlorophyll, has a maximum at a middle depth in the euphotic layer. Chlorophyll may be found below the compensation depth because the algae and the grazed material sink, the former slowly and the latter quickly. If the layer of water mixed by the wind extends below the compensation depth the algal cells are mixed downwards and suffer a loss in production. Indeed, in early spring, there is a critical depth to which

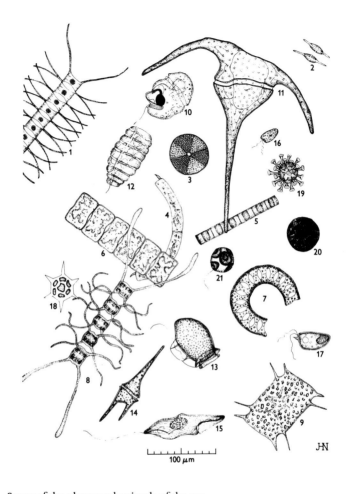

Fig. 1–3 (a) Some of the plants and animals of the sea.

1 Chaetoceros decipiens	10 Protoerythropsis vigilans	
2 Nitzschia closterium	11 Ceratium tripos	
3 Actinoptychus undulatus	12 Polykrikos schwarzi	Dinoflagellates
4 Rhizosolenia stolterfothii	13 Dinophysis acuta	
5 Skeletonema costatum — Diatoms	14 Ceratium furca	
6 Navicula membranacea	15 Gyrodinium spirale	
7 Eucampia zoodiacus	16 Rhodomonas baltica	Small flagellates
8 Chaetoceros laciniosus	17 Prorocentrum micans	
9 Biddulphia regia	18 Distephanus speculum — Silicoflagellate	
	19 Discosphaera thomsoni	
	20 Coccosphaera pelagica	Coccolithophor
	21 Coccolithus huxleyi	

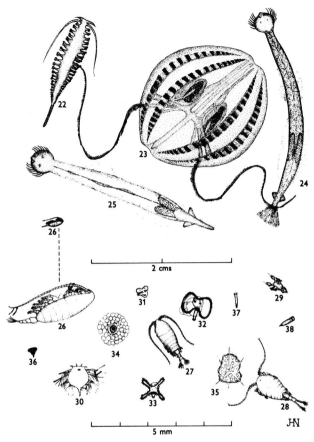

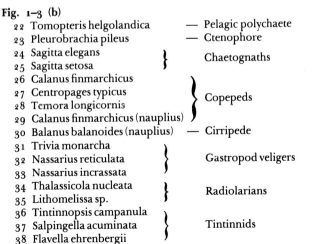

Fig. 1-3 (b)
22 Tomopteris helgolandica — Pelagic polychaete
23 Pleurobrachia pileus — Ctenophore
24 Sagitta elegans ⎫
25 Sagitta setosa ⎭ Chaetognaths
26 Calanus finmarchicus ⎫
27 Centropages typicus ⎪
28 Temora longicornis ⎬ Copepeds
29 Calanus finmarchicus (nauplius) ⎭
30 Balanus balanoides (nauplius) — Cirripede
31 Trivia monarcha ⎫
32 Nassarius reticulata ⎬ Gastropod veligers
33 Nassarius incrassata ⎭
34 Thalassicola nucleata ⎫
35 Lithomelissa sp. ⎭ Radiolarians
36 Tintinnopsis campanula ⎫
37 Salpingella acuminata ⎬ Tintinnids
38 Flavella ehrenbergii ⎭

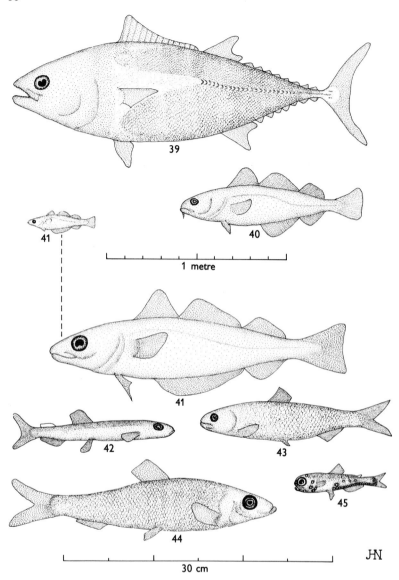

Fig. 1–3 (c)
- 39 Thunnus thynnus — Tuna (Bluefin)
- 40 Gadus morhua — Cod
- 41 Merlangius merlangus — Whiting
- 42 Mallotus villosus — Capelin
- 43 Sardina pilchardus — Pilchard
- 44 Clupea harengus — Herring
- 45 Myctophum punctatum — Myctophid

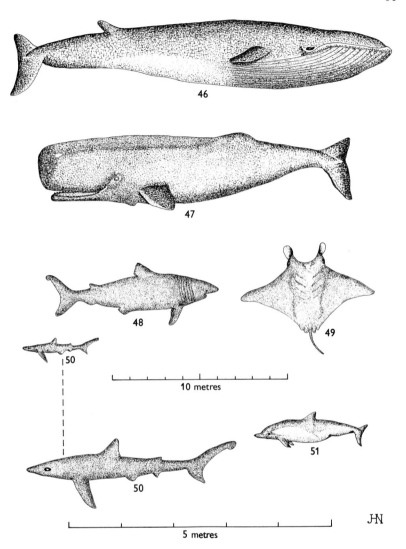

Fig. 1–3 (d)
- 46 Balaenoptera physalus — Fin whale or Razorback
- 47 Physeter catodon — Sperm whale
- 48 Cetorhinus maximus — Basking shark
- 49 Manta birostris — Manta ray
- 50 Prionace glauca — Blue shark
- 51 Delphinus delphis — Common dolphin

photosynthesis integrated in depth equals the respiration integrated in the same water column. Production cannot start until that critical depth exceeds the depth of mixing. Subsequently, as the ratio of compensation depth to depth of mixing increases, the algal reproductive rate increases.

When production starts in temperate seas, the algal reproductive rate may be as low as 0.1/d. Production increases very sharply during the spring as the winds slacken and as the sun's radiation increases; the algal reproductive rate may increase to 0.7–1.0/d. The seasonal increase in temperate waters is sometimes called the *spring outburst*, or the *production cycle*.

Nearly all the algae are eaten by the herbivores. In temperate waters an overwintering generation of adult copepods spawns when enough food is available after the spring outburst has got under way. When the new generation grows up some three weeks later it has a considerable grazing capacity and quantities of algae are eaten and so the spring outburst in chlorophyll takes the shape of a bell-shaped curve in time. There may or may not be an autumn outburst as the overwintering generation of adult copepods leaves the surface layers. Such a pattern of seasonal production follows from the complete inhibition of production during the winter. In tropical and subtropical waters, production continues throughout the year and the outburst, if it exists, has a low profile. In upwelling areas, production starts in the cool rising water at the compensation depth of 30–50 m and as the water rises at about 1 m/d, a full spring outburst occurs at the surface because, as in temperate waters, production has started from very low levels indeed. Very roughly the production in temperate seas and in upwelling areas is an order of magnitude greater than that in the tropical and subtropical deep ocean.

During the daytime, grazing by the herbivores is minimal because they live below the euphotic layer and so the algae produce as much as they can. During the evening, the animals rise towards the surface and feed on the algae. In this way is obtained the greatest production from the algae and the greatest transfer to animal flesh. In the early morning the animals descend to their day-depth where they remain till the following evening, the smaller animals in shallow layers and the larger ones deeper. The vertical migration of herbivores extends in depth to 200 m and it is accompanied by another migration of larger animals to greater depths. In the shelf areas, fish rise off the bottom at night with various bottom-living crustacea. In the deep ocean, myctophid-like fish, large prawn-like crustacea, krill and squid rise from about 400–800 m towards the surface each night and feed upon the herbivores. This mass of animals can be detected by recording echosounders and is called the *Deep Scattering Layer*.

As the animals migrate they remain in the same light environment throughout the day. There are a number of ecological consequences. First, the algae photosynthesize and reproduce in daylight with the least grazing mortality. Secondly, because the animals arrange themselves in order of depth by size, bigger ones deeper, the chance of predation in the daytime is reduced. Thirdly, the animals rise towards the surface to feed upon the algae if they are herbivores, and upon the herbivores if they are carnivores, and so on up the food chain; the transfer of material through the food levels occurs in the early morning in the near surface layers. Fourthly, material is transferred into deeper water either as assimilated food or as excretory products. Fifthly, although some material is transferred in the surface layers, much excretion occurs there and some of the nutrients are regenerated to be used again by the growing algae. The phenomenon of vertical migration plays an important part in sewing together a number of distinct ecological processes in the sea.

The system of algae, herbivores and carnivores, both primary and secondary, constitutes a food chain and any link in it is called a *trophic level* or *food level*. Because energy is lost in transfer from one food level to the next and because the animals in the higher levels are larger (and live longer), the food chains correspond to the Eltonian pyramid of numbers. There is, however, a sense in which the name 'food-chain' is a misnomer because the real structure is more like a *food web*. Few herbivores or carnivores feed on single species, but on an array, particularly as they grow. The structure of the food webs is complicated, particularly in tropical waters where there are large numbers of species to share the relatively scarce food. In the richer temperate waters and the upwelling areas in subtropical seas, where there is more food to go round, there are few species and the food webs are somewhat simpler.

The mechanisms of transfer of energy depend upon the algal production in the euphotic layer and the aggregation of all other food webs in the surface layer at night by vertical migration. In the shelf seas there is a fall out of organic material to the sea bed and from this residue of productive processes an almost autonomous food web is established, the *benthos*. Within the sediments, the protozoa (and in very shallow water, unicellular algae) constitute the microbenthos which subsist on the smallest particles and perhaps on dissolved organic material. Nematodes and small crustacea live also within the sediments and feed on the protozoa. On the surface of the sea bed live the larger epibenthic animals, polychaetes, molluscs, echinoderms and some crustacea that feed upon suspended particulate material. For fishes the benthos is particularly important because it provides food for the demersal populations such as plaice, sole, cod and haddock.

1.4 The production of life in the sea

The production of algae in terms of grams of carbon beneath one square metre can be estimated readily from radiocarbon experiments. An ampoule of sodium bicarbonate, in which the carbon is radioactive ($Na_2 H^{14}CO_3$), is added to a sample of sea water from which the animals have been filtered. For a period (for example, dawn to midday) it is exposed to the average light intensity at the depth from which it was taken; similar samples are taken from an array of depths in the euphotic layer. When the experiment is finished the samples are filtered through membrane filters and the activity is counted with a geiger or scintillation counter and it represents the amount of organic carbon synthesized. Expressed as mgC/m^3/d, it is the increment of living material in the euphotic layer and so depends a great deal on the amount of material that was there to start with. Figure 1–4 is such a chart showing the production in mgC/m^3/d throughout the world ocean.

The figure shows high production in high latitudes and the least production in the centres of the subtropical anticyclones, for example, the Sargasso Sea, and between the extremes there is a difference of about one order of magnitude. The poleward edges of the anticyclones are quite rich, as are particular areas such as the North Sea, Georges Bank and the North West Pacific gyral and, of course, the Southern Ocean. Some upwelling areas are richer than others, for example, in the California current, the Canary current (off northwest Africa), the Benguela current off South Africa and Namibia and in the Peru current. In the Indian Ocean, there are upwelling areas governed by the monsoons. During the southwest monsoon (April to September) there is upwelling off Somalia, off southern Arabia and Burma and, at the very end of the season, off the Malabar coast of India. During the northeast monsoon (October to March), there is upwelling off the Orissa coast of India and there are various minor upwellings in the Indonesian area, for example, south of Java. In the equatorial complexes there is quite high production in the currents and in the region of the undercurrent, which may explain why the Line (the equator) was favoured as a fishing ground by the old sperm whalers out of Nantucket.

The transfer from one trophic level to another, for example from algae to herbivores, is estimated by the coefficient of ecological efficiency, which is defined as the ratio of the yield from one trophic level to that from the one below, e.g. the quantity of herbivores eaten divided by the quantity of algae eaten. This efficiency has often been assumed to be about 10% between trophic levels, but it may be higher; indeed, the transfer from algae to herbivores may be as high as 15–20% in the sparse water at the centre of the subtropical anticyclones, yet in the upwelling areas it might be as low as 5%. Consequently, the production of herbivores differs between the rich and poor areas by only a factor of three or so, in contrast to the order of magnitude in primary production alone.

As might be expected the major fisheries are found in the richer areas. There are three main classes of fishery, those on the broad temperate shelves in the northern hemisphere, those in the upwelling areas, the shelves of which are usually rather narrow (except off southern or

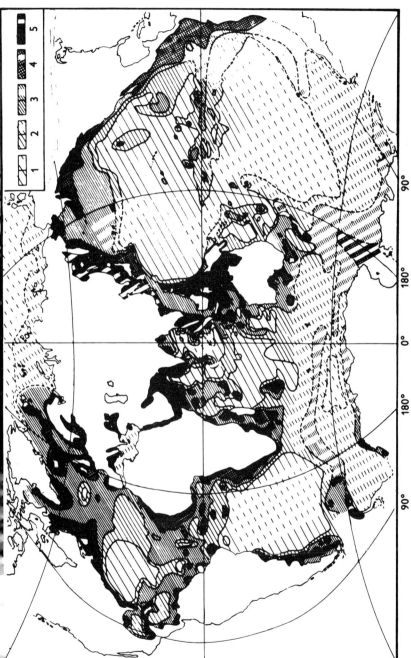

Fig. 1–4 Production in the world ocean in $mgC/m^3/d$: (1) 0–100 (2) 100–150 (3) 150–250 (4) 250–650 (5) 650+. From KOBLENTZ-MISHKE, O. I., TSVETKOVA, A. M, GROMOV, M. M. & PARAMONOVA, L. I. (1973). Primary production of the world ocean. In: *Scientific Exploration of the South Pacific*. Nat. Acad. Sci., Washington, 183–93.

southwest Africa) and those of the deep tropical ocean. At one time, people believed that the deep ocean was an immense reserve of protein. In total tonnage, it would be, but the small fish of the Deep Scattering Layer are too dispersed to be caught with present methods. The tuna-like fishes are already well exploited and it is not likely that the best yield from all the subtropical ocean will exceed five million tons. In the upwelling areas there are sardine-like fishes and hake and some of these stocks remain under-exploited, for example, hake and anchovy off California, sardines off southern Arabia and other stocks in the Indian Ocean On the temperate shelves, exploitation is extensive and probably not many stocks remain to be discovered. On a world scale there may be a potential annual yield of up to one hundred million tons, but as nearly sixty million tons are already taken each year, the prime task is conservation.

2 The Biology of Fishes

2.1 Introduction

One of the most remarkable characteristics of fishes is their high fecundity; even a small fish like a herring lays 40–60 000 eggs each year and a large cod may produce as many as ten million. Two survivors are needed to replace their parents and the enormous loss has been described as a waste in the face of a hostile environment. Predators provide enough danger, but the physical environment is benign. The loss in weight to predators of all sizes is balanced by the gain from the food chain by the two survivors as they grow. During its life a large cod may grow by seven orders of magnitude from an egg weighing 1 mg to an adult of 10 kg. The fecundity of ten million is accommodated within the large cod because the egg only reaches its full weight by absorbing much water on hatching. The eggs of nearly all bony fish are about the same size (with some exceptions such as salmon and halibut, the eggs of which are very much larger). The larvae hatch into the plankton at about 3 to 6 mm in length. They are transparent, with prominent dark eyes, a large yolk sac and they swim by wriggling through the water.

Because they live as plankton they drift in the water and make daily vertical migrations from below the euphotic layer towards the surface where they feed at dusk and at dawn. Their first food is often of large algal cells but the food on which they mainly rely comprises copepod nauplii (themselves larval stages) or, for short periods, the larval stages of molluscs. The end of larval life in the plankton comes at metamorphosis when the transparent finless animals grow fins and scales and change their shape into that of little fishes. The degree of change varies in different groups; it is a straightforward one in cod and herring, but in flatfish it is a profound change by which the animal lays on its side on the sea bed and the head is twisted to accommodate the transformation.

When fish metamorphose they are ready to settle on a nursery ground, usually a beach or a bank. The migration from spawning ground to nursery is called the *larval drift*; that of the eel extends from the Sargasso Sea, where the eggs hatch, to the mouths of European and North American rivers, where leptocephali metamorphose into elvers before they swim upstream. Juvenile flatfish and herring grow through their first one or two years on open beaches and the young cod live on offshore banks. All of them, as they grow, tend to move off slowly into somewhat deeper water and this movement has been described as a horizontal diffusion away from the shore. During this period of their lives the animals feed on local foods; herring on smaller copepods, and plaice on copepods, bivalves and worms. Their spread into deeper water gradually leads them to grounds where the adults live and they may be said then to have recruited to the adult stocks.

The commercial fish species are the abundant ones and during their adult lives, most migrate from feeding ground to spawning ground and back again every year. Pelagic fish are those, like herring or sardines, that swim in shoals in midwater and only reach the bottom now and again; the herring, for example, spawns on the sea bed. Demersal fish, like cod, haddock or plaice, are caught on the bottom but make excursions up the water column into the midwater. The pelagic fish that shoal grow by a factor of about 1.5 during their adult lives, but demersal fish during the same period of their life histories may grow by an order of magnitude. The tuna-like fishes, predators that do not shoal (or if they do, very loosely), grow considerably during their adult lives, like the demersal fish.

The spawning season may last for ten days or so, as in herring, for three months, as in cod, or for most of the year, as with some tuna. The eggs take up water as they are laid and they are spawned in batches at intervals. In temperate and high latitudes, the time between batches is short, but in low latitudes there is a tendency to extend the time between batches and such fish are called serial spawners, because the batches can be detected histologically by the size distributions of eggs in the ovary. The short period spawner may be adapted to the discontinuous production cycle of higher latitudes and the serial spawner is perhaps associated with the continuous one of lower latitudes. The connection between the production cycles and the life of fishes will be described in a later section, but the connection itself implies that the larval drift is a most important period during the life cycle of fishes.

2.2 Growth and mortality during the life cycle

Fish are among the larger animals in the sea and they live for quite a long time. In temperate waters at least, age can be determined from annual rings on the scales, as in herring or salmon, or on the otoliths (or earstones) in most other species. The Australian whitebait (*Labidesthes*) lives for a year or so, sprats for three to five years, herring for ten or twenty years, plaice for up to thirty-five years and sturgeon in Lake Winnebago in Wisconsin, U.S.A., have been caught at an age of 154. It is possible that fish in low latitudes grow faster than those nearer the poles for the bluefin tuna reaches a weight of 50 kg in three years, but may not live longer than about fifteen years. Although most fish must die by being eaten, some may well die of old age; towards the ends of their lives the myofibrils in the white muscle of cod and plaice are about half as thick as those of younger adult animals. This means that the attack or escape speed is reduced considerably; very old flatfish were once found off New-

§ 2.2 GROWTH AND MORTALITY DURING THE LIFE CYCLE

foundland with their white muscles reduced to an almost watery state and it is remarkable that they survived at all.

Larval fish grow in weight very quickly, 6%/d in plaice, 12%/d in haddock. During the period of metamorphosis growth rate is restrained partly through endocrine influences and partly through a change in feeding habits. A juvenile plaice grows from about 1 g at the end of its first summer on the sandy beaches to about 300 g when it recruits to the adult stock in deeper water offshore, three years later. The specific growth rate, which is $(W_1/W_0) \, 100$, declines with age:

$$\frac{W_1}{W_0} = e^{k't}, \text{ or } W_1/W_0 = \exp k't,$$

where W_0 is the weight at the start of the period, W_1 that at the end, k' the instantaneous growth coefficient and t the time interval in years. The decline in specific growth rate with age continues through adult life although by the age of ten the plaice may weigh as much as 1 kg. During the three years of juvenile life the average specific growth rate is 700%/yr, but in adult life it is as low as 19%.

The decline of the specific growth rate with age shows that the growth of fishes from metamorphosis onwards is an exponential function that is modified towards an asymptote at infinite age. This relationship is expressed conveniently by the von Bertalannfy growth equation:

$$W_t = W_\infty [1 - \exp - K(t_1 - t_0)]^3$$

where W_t is weight at time t, W_∞ that at an infinite age, K the rate at which growth tends towards W_∞ (and is distinct from k' the exponential growth rate) and t_0 the age at which growth starts (which may not be zero when data are graduated by the equation). The equation expresses neatly the truth that the specific growth rate declines with age until at great age it is very low indeed (Fig. 2–1).

Larval fish die almost as quickly as they grow (but not quite as much or the population will not grow in biomass). Larval plaice die at the rate of 5%/d and larval haddock at 10%/d. When the little plaice appear on the nursery beaches the death rate is 40%/month and after their first winter on the beaches it is reduced to about 10%/month. Adult plaice between the ages of five and fifteen die at 10%/year. Just as the specific growth rate declines with age, so does the specific death rate. The death rate of fishes is estimated by $N_1 = N_0 \exp -Mt$, where N_0 is the initial number in a *year class* (a brood, hatched in a given year, at any age during its life), and N_1 the number in that year class, a week, a month or a year later, or whatever is the interval time, t, and M is the instantaneous mortality rate. Numbers are estimated from ratios of stock density, the catch per unit time with a plankton net, a beach seine or a fisherman's trawl. Very little is really known about the death rates of fishes in the absence of fishing because death by fishing is often greater than the 'natural' death rate. However,

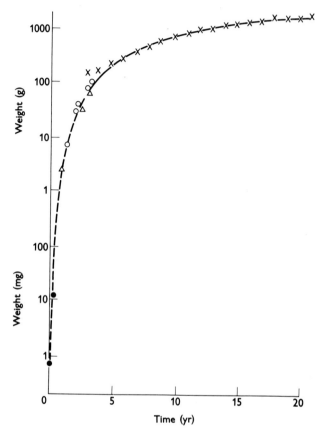

Fig. 2–1 Growth in weight of plaice fitted by the von Bertalannfy equation to older age groups indicated by (X); the other observations of younger fishes (o, Δ) and fish larvae (●) are not so fitted. (CUSHING, 1975.)

with the high fecundity of their parents, the larvae must suffer an enormous mortality, yet on the other hand if the adults live for a long time their natural death must be low. Thus, the specific death rate must decline with age as the larvae that survive predation in the early stages become themselves predators.

There may be 10^{10} fish in a large exploited stock and the biomass may amount sometimes to millions of tons. In each generation that biomass is replaced as large numbers of larvae and juveniles die and as their survivors grow. A year class that passes through the life cycle is called a *cohort* and, as the cohort grows to maturity and adult life, biomass is maximized at some middle age. This is because at any given period in middle age the specific growth rate is a little greater than the specific death rate. After middle age the reverse is true and biomass declines with age until the cohort is extinguished. In an unexploited stock the maximum biomass occurs fairly late in the life of the cohort.

The enormous loss of eggs, larvae and juveniles is not necessarily a wastage because there is a net gain of growth over death and when this is integrated in age throughout the life cycle the biomass of the parent generation is completely replaced. It represents a temporary gain from the food web, which is returned to it when the cohort is extinguished. The fishes move up the food web as they grow, eating larger prey and subject to larger predators. As biomass is maximized in each cohort the food web is exploited to the full.

2.3 The circuit of migration and reproductive isolation

The circuit of migration may be represented as a triangle with the larval drift, from spawning ground to nursery ground, forming the base and the feeding ground, the apex. The juveniles move slowly from the nursery to the feeding ground where they recruit to the adult stock. Each year the adults migrate from feeding ground to spawning ground and back again. From the spawning ground both larvae and spent fish drift away in a regular current, a downstream or denatant movement. Consequently the spawning migration is upstream or contranatant.

In temperate waters, at least, fish tend to spawn on grounds of fixed position at the same season each year. The plaice spawn in the southern North Sea between the Thames and the Rhine and this position has not changed since the area was first surveyed in 1911. In roughly the same area and in the eastern English Channel, the East Anglian herring spawn on six restricted gravel sites each year, one of which is 2 km in length by 0.5 km in breadth; there is a similar small ground, also of gravel, on Ballantrae Bank off the coast of Ayrshire, where the Clyde herring spawn. Cod have been caught in spawning condition in the Vestfjord since the twelfth century and since 1955 the spawning layers have been charted with echosounders; they are usually about 32 km long by about 3 km wide. Such fixed spawning grounds seem to be restricted to temperate waters. Tuna spawn all over the North Pacific subtropical ocean; sardines off California appear to choose precise positions for spawning but may gather where the water upwells, which might vary in position from year to year.

If fish do spawn at the same position each year in a regular current, then the larvae drift along a fixed course to a nursery ground which is the same from year to year. From the spawning ground between the Thames and the Rhine, larval plaice drift in the clearer Channel water to the northeast and as they get older they gradually sink deeper in the water column towards the sea bed. Off Texel Island, northern Holland, the tides run parallel to the boundary between stratified coastal water and fully mixed water offshore. Under this condition at certain states of the tide, the water moves inshore near the bottom and offshore near the surface. The metamorphosing plaice larvae in the lower half of the water column find themselves drifted on to their nursery ground on the Waddensee, the broad shallow flats inside the Friesian Islands. The whole larval drift takes place in the same geographical position from year to year and this depends upon the constant position of the spawning ground.

The larval drift occurs every year, but only once in the life history of the cohort. The spawning migration also takes place each year but in successive years for a particular adult cohort, as part of the whole spawning stock which comprises many cohorts. The range of the spawning migration is considerable. Even little fish like sprats migrate up to 80 or 160 km a year. The East Anglian herring spawn in the southern North Sea and eastern English Channel, yet the feeding grounds in the central North Sea and in the northern North Sea are up to 640 km distant, a round trip of at the most 1280 km a year. The Atlanto-Scandian stock of herring travels all round the Norwegian Sea, from the Norwegian coast where they spawn between Haugesund and the Lofoten Islands to the Polar Front where they feed between Jan Mayen and Iceland and by the East Icelandic current north of the Faroe Islands back to the Norwegian coast, an annual distance of perhaps 1900 km. The Arcto-Norwegian cod that spawn in the Vestfjord feed on the banks near Bear Island, off West Spitzbergen and on the banks east of the North Cape as far as Novaya Zemlaya; their annual travels may amount to 2200 km or more. Albacore tagged off southern California have been recovered a year or so later off Japan and bluefin tuna tagged off Florida were recaptured eighteen months later off Norway. Such transoceanic migrations are probably much longer than the straight course between the point of liberation and that of recapture if the fish move in the direction of the major currents. The spawning migration of big fish is longer than that of little fish and the distances covered are considerable.

Fish stocks appear to be contained within oceanographic structures. Tuna live mainly in the subtropical circulations; the yellowfin tuna spawn rather more between the Central Pacific and the Philippines than elsewhere in the North Pacific and the juveniles tend to be found in the Kuroshio extension drifting eastwards at 40° N; the stock is contained within the whole circulation of the subtropical North Pacific. The Atlanto-Scandian herring are contained within the Norwegian Sea and move in the same direction as the main current system; the northern and central North Sea herring migrate round the North Sea in the same direction as the main North Sea swirl. Spent cod from the Vestfjord drift in the West Spitzbergen current or the North Cape current till they reach the banks where they feed. On the spawning migration the cod either swim against the main currents or drift with the countercurrent and leave it as they pass the Lofoten Islands and turn into the Vestfjord. The fish make use of the current system to reach certain positions at which they leave it. In the same way, the Atlanto-Scandian herring migrate in the East Icelandic current from Iceland to north of the Faroe Islands and cross beneath the Atlantic stream to reach the Norwegian coast at the right time and at the right position.

Although the spawning season is rather long, up to three months in some species, its peak date appears to be fixed. For plaice in the southern North Sea, the Atlanto-Scandian herring, the Arcto-Norwegian cod and for the sockeye salmon of the Fraser River system in British Columbia, the standard deviation of the peak date of spawning is less than a week. They are all temperate species; subtropical ones such as the Californian sardine spawn at variable seasons and there does not really appear to be a peak date, let alone a regular one. If spawning takes place at the same time each year and the larvae drift in a regular current, the nursery ground is occupied at the same time each year and is pre-empted from any competitor of the same species. Because there is a fixed term to larval development, the nursery ground must lie within a defined distance of the spawning ground. Plaice larvae from the Moray Firth might reach the Southern Bight of the North Sea if the current system were the only determinant, but at the known speed of the current the larvae must metamorphose long before they could reach the southern North Sea. Thus, reproductive isolation has been obtained.

If reproductive isolation has been established it should be identifiable in genetic terms. Groups of the North Atlantic cod have been known for a long time, for example, Grand Banks, Labrador, Iceland, Faroe, Barents Sea, North Sea. Some blood proteins are genetically determined, for example, haemoglobins and transferrins and they can be distinguished by electrophoresis, i.e. they move at different speeds across a substrate in an electric field. Two haemoglobins and seven transferrins have been distinguished in the North Atlantic cod and any stock is shown to be separate from its neighbour; the chance of mixture between such groups is as low as 1 in 10^4.

2.4 The generation of the recruiting year classes

Most fish in temperate waters are spring or autumn spawners, although summer spawners do exist in some areas. The spring outburst, or production cycle, varies in time of onset, in magnitude and duration. Because there are usually few copepod nauplii available before production starts, fish larvae must in general starve if they are hatched before then. Hence, it is not surprising that fish tend to spawn in temperate waters at such a time that their larvae hatch when the spring outburst is under way. If production continues all round the year in tropical and subtropical waters, there is no need for a fixed spawning season and, as we have seen, it is absent. In the upwelling areas, fish may spawn during the season of upwelling and so their eggs hatch into food-rich water. Thus, in one way or another fish spawn at such a time to let their larvae feed well.

In temperate waters, fish spawn at a fixed season whereas the timing of onset and of peak production in the spring outburst is variable. Hence the production of larvae is matched or mismatched to that of their food (Fig. 2-2). For example, during the cold winter of 1962–3, the water in the Southern Bight was cold and the development of larval plaice was delayed. Such cold winters that occur every decade or so are due to a persistent anticyclone over southern Scandinavia that brings very cold easterly winds to southern England; such winds do not mix the water so much as the northerlies and the spring outburst might have been advanced in the favourable marine weather. The chance of the match of larval production to that of their food was therefore high and the 1963 year class of plaice was the highest since records started in the late twenties.

The variability of year class strength is high. For the East Anglian herring it lies between a factor of three and ten. That of the Arcto-Norwegian cod ranges over about an order of magnitude (Fig. 4.2). For the Atlanto-Scandian herring and the North Sea haddock, recruitment varies by two orders of magnitude and that of the North Sea sole varies by a factor of sixty. Such series have been examined for periods of twenty to eighty years. In the case of some cod-like fishes, there is no upward or downward trend with time. However, the 1962 year class of haddock in the North Sea was twenty-five times larger than any predecessor and those of 1967 and 1972 were also very large. A profoundly different form of variability is shown in the alternation between the fishery for the Atlanto-Scandian herring and that for herring off the Bohuslån coast of Sweden every 50–70 years; during the period of abundance of one, the fishery for the other is extinct. Thus there are three forms of variability: that about a steady mean level, that about a slow upward or downward trend with time and the periodic dramatic changes that may generate a large fishery or extinguish it.

The variation in the onset of the spring outburst in temperate waters depends upon the sunshine and upon the stress of wind and its direction; the direction is important because if the wind has blown over the ocean for great distances it generates heavier waves than if it blew off the land. Slight shifts in wind direction near a coastline at the same strength can cause considerable differences in vertical mixing. In the shelf seas the depth of mixing is usually wind driven and, as the wind tends to slacken as winter passes into spring and early summer, the depth of mixing lessens. At the same time, as the sun's angle rises and the length of day increases, the compensation depth increases during the season and so the production ratio (compensation depth/depth of mixing) increases. Thus there are good climatic reasons for the variation in the onset of the production cycle and, of the two, the change in wind stress from year to

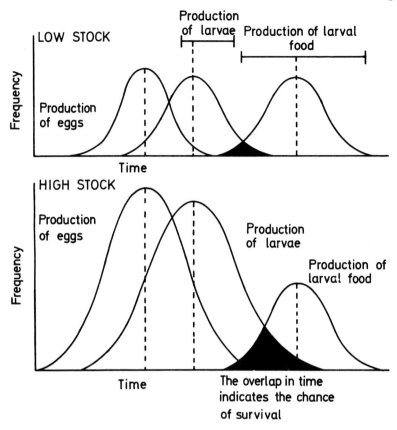

Fig. 2–2 The match of larval production to that of their food: the figures show that there is a greater chance of match at high stock, i.e. that recruitment becomes less variable. (CUSHING, D. H. (1976). *Geog. J.*, **142**, II, 224.)

year is the more important; on a longer time scale, changes in wind direction as the result of climatic change may be important. Many attempts have been made to relate the magnitude of year classes with wind strength, because the wind was supposed to blow the larvae away from productive areas. However, in the course of decades, the direction changes. The match/mismatch thesis requires that wind stress be estimated including the effects of directional change.

There are two observations that indicate a relationship with climatic events. For nearly thirty years the year class strength of cod and haddock in the different North Atlantic stocks were classed in five groups, from 'very poor' to 'very good'. In some particular years, for example 1950, all year classes throughout the ocean were classified as 'very good' and there were other years when all the year classes were 'poor' or 'very poor'. Such a correspondence across the whole ocean can well be explained by a factor as pervasive as climate. The second observation is a good correlation between year class strength of the Arcto-Norwegian cod and the widths on pine trees in the same region for a period of fifty-five years. Although this statement somewhat simplifies the actual procedure used, the only factor common to the two data series is a climatic one. Hence it is reasonable to suggest that the magnitude of recruitment is responsive to climatic factors as mediated in the match of larval production to that of their food.

Although the weather is notoriously variable, particularly in the British islands, climate is less so. For example, since the first decade of the century it became gradually warmer in the British Isles, reaching a peak in the thirties and early forties and since then it has become somewhat cooler. The sea in the northeast Atlantic became warmer by as much as 1°C or so, which represents an enormous increment of heat. Perhaps the most spectacular event during the period of warming was the colonization of West Greenland banks by cod from Iceland. There were no cod on the offshore banks at West Greenland in the first decade of the century and it is possible that recruitment to that stock has failed since 1968.

Because the long term changes seem to vary in a number of decades, there is not enough information to establish their source. However, in British waters, the prevailing wind shifts in some decades from westerly to southerly and back again. The period of this shift appears to be the same as that of the alternating fisheries of the Atlanto-Scandian and the Bohuslån herring. The evidence for the alternation goes back for centuries. However, the temporal changes in wind strength and direction might account for the slow trends in recruitment that occur in some stocks of fish. An important character of the wind at sea is the distance across which it blows, or the fetch, and the great oceanic swell is generated by the long fetch; then a change in direction from westerly to southerly along the Norwegian coast (that runs from southwest to northeast) would reduce the sea state considerably, reduce the depth of mixing and advance the time of onset of production. Such is the form of explanation that might account for the appearance and disappearance of the Atlanto-Scandian herring fishery.

In an earlier section, it was noted that as a cohort proceeds through its life history its biomass is maximized by the countervailing forces of growth and death. Year class strength is highly variable and if it is modulated in the way suggested, then the variability originates at an early stage in the life cycle. Indeed, in some stocks, the year class strength can be forecast in the first year of life. Each cohort takes from the environment all it can, starting with the match of larval production to that of its food and continuing with the maximization of the biomass. In any stock, there are many age groups and so it represents the average of many cohorts. The cohort has an exploratory function, in assessment and extraction from the environment; the stock has a conservative function by which the quantity of eggs produced in any one year is effectively the average of a number of cohorts, the average of a number of explorations of an environment that changes all the time.

3 The Population Dynamics of Fishes

3.1 Introduction

When fishermen's catches are landed on the quay, they are grouped in sizes, if need be, and by vessels, for the auction. The total catch for the voyage may be laid out by size groups. The catch per day, or the catch per 100 hours fishing, is called the *catch per unit of effort*, the *average catch* or the *stock density*. *Fishing effort* is time spent fishing, or time to kill fish. The mate of the fishing vessel is asked how much fish was caught on which ground and how many hours were spent fishing. The collector of statistics then has all the information needed to establish the catch per unit of effort for a particular ground. Such information is pooled between ships and ports and is grouped by species, by month and statistical square (thirty by thirty nautical miles). Distributions of stock density, of total catch and of fishing effort (in hours' fishing) yield many of the facts needed by the fisheries' biologists in their work.

On the markets, scientists measure the fish in length and take samples of scales or otoliths. In the laboratory such samples are used for age determination; by combining age and length measurements from each fish, a table is made, of age in rows and lengths in columns, by which length distributions may be converted into age distributions. As the samples measured were taken from known vessels, the age distribution can be expressed in stock density, e.g. so many five-year-olds per 100 hours fishing, etc. Such distributions are grouped by months or quarters and by areas and are combined to give a single age distribution for each year.

The age distributions are used in two ways. First, the mean lengths at each age can be extracted and from such estimates a growth curve can be constructed from the von Bertalannfy growth equation; fish can be measured in length more quickly than they can be weighed and the lengths may be converted to weight with a weight/length relationship. Secondly, an estimate of mortality is made from the age distributions in numbers because the three-year-olds in one year belong to the same year class as the four-year-olds in the next. The instantaneous coefficient of mortality from year to year, Z, is derived from $N_2/N_1 = exp(-Z)$, where N_1 is the number in the first year and N_2 the number in the second; then $ln(N_2/N_1) = -Z$ where ln is a napierian logarithm. The percentage mortality is

$$\frac{(N_1 - N_2) 100}{N_1} = \left(1 - \left(\frac{N_2}{N_1}\right)\right) 100 = (1 - \exp(-Z)) 100$$

Such is the information taken from the fish markets. It is of little use, however, unless incorporated in some model of the fish population. The first descriptions of fish populations were made around the turn of the century, but they were little more than qualitative models of the dynamic processes. BARANOV (1918) stated the problem, but his work was not noticed until after the Second World War. The first model to put the problem fully, and at the same time simply, was that of RUSSELL (1931):

$$P_2 - P_1 = R + G - F' - M'$$

P_1 and P_2 are stocks in year 1 and year 2, R is the annual increment in recruitment, G is the annual increment in growth, F' is the annual decrement due to fishing mortality, and M' is the annual decrement due to natural mortality. From the age distributions in stock density, estimates of each parameter as instantaneous coefficients can be obtained. The next most important problem was to separate total mortality, Z, into its natural, M, and fishing, F, components.

3.2 Estimation of fishing mortality

There are many historical records of fish having been tagged to identify them, but PETERSEN (1894) was the first to use them for the quantitative purposes of fisheries research. The Petersen tag is a numbered button attached to the fish with a silver wire and it can be identified by its number on recapture. As might be expected, tags have evolved in divergent directions for different conditions. They can be shot into the body cavity for recovery in fish meal factories with magnets. Some are neutrally buoyant. Others are little plastic flags. Among specialized ones are very small transponders that emit a signal when stimulated acoustically; with such tags, individual fish can be tracked in detail for days at a time. Many of the most spectacular results in fisheries biology have emerged from tagging experiments, for example, the transoceanic migrations of tuna or the return of Pacific salmon to their parent streams.

One of the most important uses of tagging experiments is to estimate fishing mortality, based on the relationship, $C = FP$, where C is catch; consequently, $F = C/P$. A stock of tagged fish is released into the sea and is assumed to represent the natural stock; fishing mortality is estimated from the ratio of the catch of tagged fish to the stock of tagged fish. If 30% of tagged fish are recovered within a year, that is the annual percentage mortality. This percentage is

$$\frac{(N_0 - N_1) 100}{N_0} = \frac{(N_0 - N_0 \exp -(F + X)) 100}{N_0} = (1 - \exp -(F + X)) 100$$

where X is the instantaneous other loss coefficient. Stock (in numbers), P, may be estimated from $P = N_m (Y_n + 1)/(N_r + 1)$, where N_m is the number marked, N_r the number recaptured and Y_n the catch in numbers.

Such a simple estimate can be made with a single experiment in a lake if most of the recaptures are taken quite quickly; the object of such an experiment is to estimate numbers in the stock. In the sea, because fish migrate from one fishing ground to another and because stocks are widespread, recoveries are recorded for a number of years; the aim is to estimate fishing mortality. The apparent loss rate in time estimates total mortality. However, some biases have to be eliminated first: the immediate death rate due to tagging (which can be estimated by keeping tagged fish in a tank on deck for a time), the indirect death rate because predators may notice a conspicuous tag, and the failure to recover a tagged fish from the catch. There is a body of statistical methods available to estimate errors once the biases have been eliminated. However, a tagging experiment at sea suffers from one main disadvantage: it may be a long time before the stock of tagged fish is properly mixed with the natural one and by that time the proportion of recoveries may be low. Consequently, tagging experiments at sea tend to be used in a restricted manner, for example, at the start of an investigation when no information is available from other sources.

Once an estimate of fishing mortality is available it is related to fishing effort, the time spent fishing; $F=qf$, where q is the coefficient of catchability (for example, 0.001 F per hundred hours trawling) and f is fishing intensity, or fishing effort per unit area. If fishing intensity has varied considerably during a period, total mortality, Z, can be plotted on fishing intensity; the slope of the relationship estimates q and the intercept on the ordinate estimates M, the instantaneous coefficient of natural mortality. Because $P=C/F=C/qf$, then $q=C/Pf$; in other words, because catch per effort is a proper index of stock, the catchability coefficient is the ratio of stock density (C/f) to stock, P. The value of this conclusion will emerge in the next section on the Graham-Schaefer model.

Age distributions are expressed in numbers of stock density. Let R be the youngest fully recruited age group in a year class in numbers or numbers per unit of effort at the beginning of the year. At the end of that year R is reduced to R exp $(-Z)$ by fishing and natural mortality. Then the number of deaths during that year is $R-R\exp(-Z)=R(1-\exp(-Z))$. The proportion of deaths due to fishing in that year is (F/Z) and the catch in numbers, C, is given by $C=(F/Z)R(1-\exp(-Z))$. Such quantities can be added in all age groups in one year or in a number of years within a year class, i.e. $C=(F/Z)R(1-\exp(-Z\lambda))$, where λ is the number of years exploited within the year class. If the system is in a steady state, this catch equation by age groups in one year is equivalent to that for a year class throughout its life in the fishery.

The use of the catch equation by age groups within a year class can be extended to make estimates of stock and of fishing mortality. The equation may be written most simply as $C=EN$, where E is the exploitation rate $(=(F/Z)(1-\exp(-Z)))$ and where N is stock in numbers. In the last age of the year class, λ, $C_\lambda = E_\lambda N_\lambda$; Z_λ comprises an average value of M and an educated guess for F.

Then $N_{\lambda-1} = N_\lambda \exp Z_\lambda$ and $E_{\lambda-1} = C_{\lambda-1}/N_\lambda \exp Z_\lambda$. From this estimate of exploitation rate, $F_{\lambda-1}$ can be derived and the process can be continued back through the year class to the youngest age group, the recruits. This method of cohort analysis yields a matrix of estimates of stock in numbers by ages and by calendar years and another of estimates of fishing mortality.

In earlier years it was believed that the natural mortality rate was high partly because people were impressed by the fact that fish in the sea die by being eaten when fishermen were absent and partly because they disliked any restraint to fishing activity. Today the one or two good estimates of natural mortality have led us to believe that it is relatively low, which of course is reasonable if fish live for one, two or three decades.

3.3 The Graham-Schaefer model

The simplest model of a fish population states that catch is a fraction of stock and that it is sustained because the ratio of recruitment to stock rises as the rate of fishing increases. It is assumed that as stock declines there is no absolute reduction of recruitment. The evidence for this was merely that some fisheries had lasted for centuries. W. F. Thompson related the stock density (catch per *skate*, or catch per *set* of six hooks linked on a single branch line) in the Pacific halibut fishery to the fishing effort (the total number of hooks) and the curve in Figure 3–1 shows that stock density declines as fishing effort increases. Thompson's rule had been stated earlier but this was the first explicit formulation. Thompson used the relationship to regulate the Pacific halibut fishery merely by reducing fishing effort by small steps and so increasing the stock density. With simple models he showed that catch would increase at the same time, which it did.

The first model to put catch in the same terms as catch per unit effort was that of Graham in his use of the logistic curve. This equation states that a population increases to a saturation level; its net rate of increase (or recruitment per unit stock) declines to zero when the population matches the carrying capacity of the environment. Then, in an exploited stock, with stock reduced as catches increased, the recruitment per unit stock should increase to compensate for loss of catch. GRAHAM (1935) assumed that the net rate of increase was proportional to fishing mortality. The latter was estimated by tagging (GRAHAM, 1938). Given such a relationship, stock and catch could be estimated.

The present form of this model is that of SCHAEFER (1954). His form of the logistic curve stated that the rate of change of biomass, P, was a function of biomass up to a limiting value, P_m:

$$dP/dt = aP(P_m - P)$$

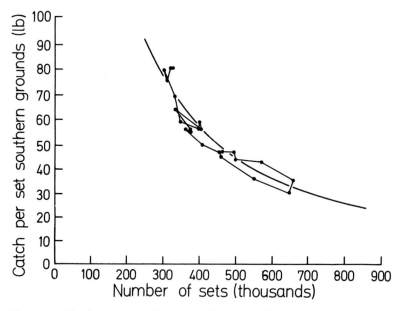

Fig. 3–1 The dependence of catch per effort, or stock density, on fishing, a rule established by W. F. Thompson. (From THOMPSON, W. F. (1950). *The Effect of Fishing on Stocks of Halibut in the Pacific*, University of Washington Press, Seattle, U.S.A.)

where a is a constant. When the stock is exploited, $dp/dt = aP(P_m - P) - FP$. Stock is estimated from the sum of catch and stock increment in the year. Thompson's rule is approximated linearly by $P = (aP_m - F)/a = P_m - F/a$. As $P = C/fq$, the rule is put in terms of stock density, C/f. If the equation is multiplied through by F, a parabolic yield equation is obtained:

$$FP = Y = FP_m - F^2/a$$

The maximum of the curve of catch on stock is defined as the *maximum sustainable yield*. SCHAEFER (1957) calculated such a yield with confidence limits for the yellowfin tuna stock in the eastern tropical Pacific. A useful rule of thumb with this formulation is that the greatest catch is taken at half the unexploited stock. Figure 3–2 shows the most famous modification of this method in which the sharp decline of catches of Antarctic blue whales falls as stock decreases. The stock had been fished so hard that it could not replace itself with the annually recruiting year classes.

The great advantage of the Graham-Schaefer model is that the only information needed is of catch and of fishing effort. The differences in stock from year to year summarize all the changes in the vital parameters given in the Russell equation. Further, if one is modified in more subtle

§ 3.4 THE YIELD PER RECRUIT MODEL

ways, for example, in an unspecified density dependent manner, it is accounted for in the difference of catch from one year to the next. There are two disadvantages: first, the causes of any change in stock cannot be ascribed and, secondly, the curve needs up to twenty observations of widely differing stock values to estimate the maximum sustainable yield. It is then improved only by the accretion of single observations from year to year.

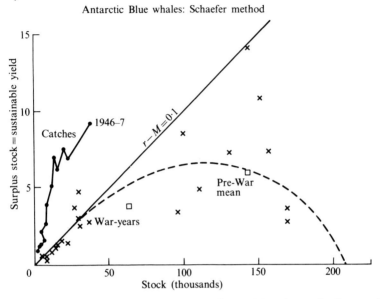

Fig. 3-2 Application of the Graham-Schaefer model to the stock of Antarctic Blue whales; the decline in stock was due to the fact that the catches were greater than could be sustained by the effective reproductive rate (reproduction less natural mortality). (*IWC Annual Report*, **14**, 67.)

3.4 The yield per recruit model

This model, due to BEVERTON and HOLT (1957), originated partly in the Russell equation and partly in the methods of operational research developed during the Second World War. The basic premiss is that of the Graham-Schaefer model, $C = FP$, but was expressed:

$$Y_w = \int_{t_c}^{t_\lambda} F_t N_t W_t \, dt \qquad (1)$$

where Y_w is the yield in weight, t_c is the age at first capture in a cohort, and t is the age of extinction of the cohort. Between the ages t_c and t_λ, the life of the cohort in the fishery, numbers and weight can be integrated in time separately. To do this, the trend in numbers must be known e.g. $N_t = R \exp-(F+M)t$, where t is any age after t_c. Further, the weight at any age must be calculable, e.g. $W_t = W_\infty (1-\exp-K(t_c-t_0))^3$. Before equation (1) can be integrated, a number of constants should be determined, R, F, M, K, W_∞ and t_0. The expression is integrated in two steps, first in numbers and then in biomass, both through the year class. The expression for yield, Y, is:

$$Y_w = FRW_\infty \left[\frac{1}{F+M} - \frac{3\exp(-K(t_c-t_0))}{F+M+K} + \frac{3\exp(-2K(t_c-t_0))}{F+M+2K} - \frac{\exp(-3K(t_c-t_0))}{F+M+3K} \right] \quad (2)$$

If equation 2 is expressed as Y_w/R, there is no need to estimate recruitment. There are two reasons for this step: (a) because recruitment is highly variable, it is difficult to estimate an average unless many observations are available, and (b) recruitment in absolute numbers was assumed to be independent of parent stock at most levels of exploitation. At low stock levels, recruitment must be reduced absolutely, indeed to zero at zero stock, but they were considered to be below the range of fishing activity. We shall return to these assumptions in the next section.

Figure 3–3 shows a yield per recruit curve for the southern North Sea plaice. By comparing numbers in cohorts before and after the Second World War, Beverton and Holt were able to make one of the few reliable estimates of natural mortality, $M = 0.1$; $F = Z - 0.1$. All the constants in the yield equation were estimated from age distributions based on the market statistics. There is a maximum in Y_w/R at $F = 0.25$ and the fishing mortality before the Second World War was shown to be 0.73. A gain in total catch of about 20% could be obtained by reducing fishing by a factor of three, i.e. putting two-thirds of the fishermen out of work. This was obviously undesirable and it was shown that the same result could be obtained by raising the age at first capture, i.e. by enlarging the meshes of the trawls to let the little fish escape and put on more weight before they were finally caught. As fishing mortality would be reduced stock density would increase, by Thompson's rule, in this example by a factor of 3 and profits would also. As will be shown in the next chapter, this solution formed the basis of the international regulation in the North Atlantic that was established in the fifties.

The scientific advice in the yield per recruit model was the exploitation of information on growth and mortality available in the age distributions. Hence the model is a synthesis of the observed parameters and it depends on how well they are observed and on the population not being governed by any factor outside the system, for example, any trend of recruitment in parent stock. The model also provided an advance in management because advice on regulation could be given immediately the parameters

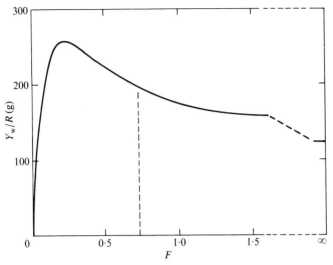

Fig. 3-3 The yield per recruit model applied to the southern North Sea plaice for the pre-war years. The dotted ordinate represents the pre-war fishing mortality; to gain an increase in catch of about 20% a reduction of fishing of a factor of three would have been needed. (BEVERTON and HOLT, 1957.) (British Crown Copyright. Reproduced with the permission of HMSO.)

were determined. Although some years of observation are needed to establish the age distribution in numbers and in weight, once established there is no need to await the slow accretion of observations to estimate the maximum yield as in the Graham-Schaefer model.

3.5 Recruitment and parent stock

Because recruitment is very variable, the plot of recruitment on parent stock may conceal the true relationship. Then independence of recruitment from stock might be assumed unwarrantably. The collapse of the East Anglian herring fishery in the late fifties and early sixties was due to recruitment failure under the pressure of heavy fishing. Many stocks of pelagic fishes have collapsed in such a way but it now seems possible that the populations of the very fecund fishes like cod can resist recruitment failure more easily than the less fecund ones like herring. The more fecund ones release more larvae into the sea and the density dependent processes may be more severe. Consequently, the stock is more tightly controlled with respect to the food available to it during its life history. Conversely, stocks of herring-like fishes are more vulnerable to collapse whether due to fishing or to natural causes.

It is generally conceded that animal populations retain their stability in numbers by density dependent processes in growth and/or mortality.

Amongst adults density dependent changes in weight could generate differences in fecundity and hence population numbers. In fishes, however, density dependent growth appears to be more or less absent amongst the adults although prominent amongst juveniles; fishery biologists, however, work with exploited stocks in which the density dependent processes may be reduced. Nevertheless, on present evidence, the population is controlled by density dependent processes during larval and juvenile life.

The mortality from egg to recruiting adult may be considered as being partly density dependent and partly density independent. RICKER's (1958) equation states:

$$R = AP \exp(-BP),$$

where R is recruitment in numbers (at the age of recruitment to the adult stock), P is stock, A is the coefficient of density independent mortality, and B is the coefficient of density dependent mortality. If density dependent mortality is low, the curve is lightly convex, and if high, the curve forms a dome; density independent mortality is the slope of the curve at zero stock. Figure 3–4 shows a stock/recruitment curve for the Arcto-Norwegian cod stock. Thus the assumption, implicit in the yield per

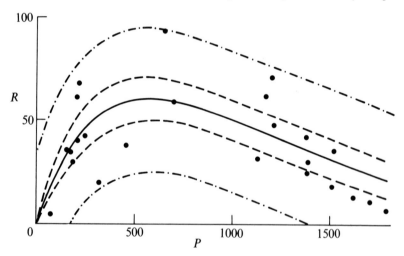

Fig. 3–4 The dependence of recruitment on parent stock in the Arcto-Norwegian cod stock; the errors to the curve are given by the pecked lines. An estimate of the error of residuals is given by the pecked and dotted lines.

recruit curve, that recruitment is independent of parent stock is not always valid, but the course of the stock/recruitment curve through the scattered points shows where the yield per recruit one can be used.

A manager would try to exploit the stock somewhere near the point of maximum recruitment. Sometimes the stock that yields the greatest recruitment per unit stock is nearly the same as that for the yield per recruit curve, but under some conditions it is less (particularly for the herring-like fishes). A decade or so ago, one of the problems for a fisheries biologist was to distinguish natural and man-made causes in a failure of recruitment. It is now clear that if a stock is maintained somewhere near the maximum recruitment per unit stock, recruitment failure due to fishing is very unlikely; further, if it failed due to natural changes, then the best exploitation had been obtained.

3.6 Conclusion

The study of fish population dynamics has evolved considerably since the thirties. Then the basic principles were established in the Russell equation and in the Graham-Schaefer model. The two major advances since that time have been the exploitation of information on growth and mortality in the age distribution by the yield per recruit model and then the formulation of the dependence of recruitment on parent stock. An advance has been made from a descriptive model to analytic ones. However, the study of the stock/recruitment problem is still hampered by the slow accumulation of annual observations; a truly analytic model that is independent of the high variability of recruitment is not yet available.

4 The International Management of Fisheries

4.1 The early history of international management

In 1899, a meeting was held in Christiania (Oslo) to consider the problems of fisheries, in rather broad terms. In 1902, the International Council for the Exploration of the Sea was founded and was based in Copenhagen, where it still sits. There were three main committees: on hydrography, on migration and on overfishing. The Council's first task was to explore the North Sea with research trawlers from a number of countries. At this time, the British interest centred on the small plaice problem; the British fleet worked close to the European coasts where small plaice lived and in summer the fishermen discarded six times as many as they retained. During the international survey, the variability of catches from haul to haul was so high that the material could not then be manipulated. This particular problem was not overcome until the twenties when the statistics of small samples were established. The solution proposed for the small plaice problem was to transplant fish *en masse* from the poor and crowded coastal grounds to the apparently richer ones on and around the Dogger Bank. This proposal was not accepted and some fifty years later it was shown that the value that might have been gained in growth would not have offset the costs of transplantation.

It is useful to distinguish between *growth overfishing* and *recruitment overfishing*; PETERSEN (1894) formulated the latter as a reduction in stock caused by recruitment failure due to heavy fishing, which the fishermen have called 'an interference with breeding'. Petersen also noticed that growth overfishing occurred when the little fish were caught without having had the chance to put on weight; such an effect could occur before recruitment was reduced. Petersen also thought that this problem could be cured by the fishermen themselves restricting their activities to the larger fish. The small plaice problem was that of growth overfishing and the yield per recruit model provides an explicit solution. The Graham-Schaefer model was applied to the plaice stock in the southern North Sea and it would have prevented either growth or recruitment overfishing. The proposal in the first decade of the present century to solve the small plaice problem by transplantation was really a solution to the problem of growth overfishing. The early models implied that the catch each year was balanced by an annual recruitment which remained about the same even if the stock was reduced by exploitation; hence the real problem was to maximize growth in the face of loss of numbers.

During the second decade of this century the Pacific halibut fishery had declined to a considerable degree. Catch per skate, the stock density, fell by a factor of six off British Columbia and total catch declined by a factor of three. A treaty was signed between the United States and Canada and as a result the International Halibut Commission was established in 1924 with W. F. Thompson as its first director. By this time fishermen were working off Alaska as well as off British Columbia; tagging experiments, morphometric measurements and egg surveys showed that the two stocks were distinct. Fish tagged off British Columbia moved tens of miles, whereas those off Alaska spread hundreds of miles from the point of liberation. Because the two regions are up to a thousand miles apart, tagged fish were not exchanged between the two stocks. With the use of Thompson's rule relating catch per unit of effort to effort, it was suggested that a relative restriction of fishing effort should improve the state of stocks. In 1930 a closed season was imposed and the number of fishermen was restrained by the issue of licences for fishing.

In the following years, stock density (as catch per skate) increased and so did catch. Figure 4–1 shows how recovery started immediately after the restrictions were imposed and how this continued for two decades in stock density and in catch. It was the first international control of a fishery and as the figure shows it was a successful one. With hindsight the control may well have been introduced with a rich year class which caused the immediate recovery; the real one was sustained in later years. Later it appeared that success was won at an economic cost; for example, in order to augment stock density in the later years the closed season was extended very considerably, which meant that fishermen, gear and cold stores were idle too long. Indeed at the worst point the fishermen worked for only a few weeks each year. Historically, the action of the International Halibut Commission was important because it was shown that a fish stock could recover if action was taken internationally to restrain the fishing effort exerted. In marked contrast the International Council for the Exploration of the Sea failed to take the same steps to solve the small plaice problem, because of the high haul-to-haul variation. When Thompson's success became clear during the mid-thirties, the Council returned to the problem and laid the foundations for the present successful control of fisheries in the North Atlantic.

4.2 The International Council during the thirties

Application of Thompson's rule to the plaice stocks in the Baltic demonstrated that stock density and catches were declining during the twenties. Through the Danish Foreign Office, the International Council was able to bring the Baltic countries together and they agreed to establish a Convention in 1933 by which plaice fishing was restricted. It was the second international agreement to regulate a fishery.

During the early thirties a number of important developments took place. The first was the publication of the Russell equation which

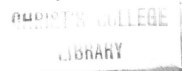

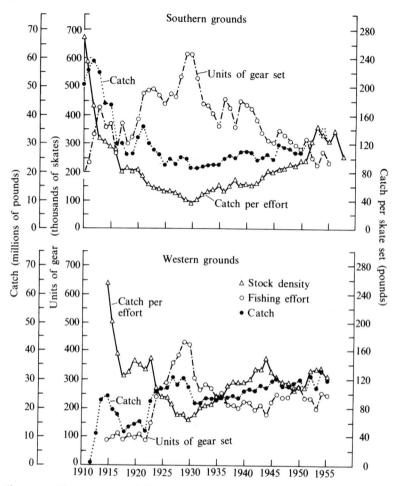

Fig. 4-1 The recovery of the stock of Pacific halibut on the southern and western grounds. (FUKUDA, 1962.)

displayed the essential processes of fish population dynamics in a simple manner. The second was the publication of a model of whale populations (HJORT, JAHN and OTTESTAD, 1933) which described the development of the stock in a cohort and how catch can be derived from it. The third event was Graham's application of the logistic curve to fish populations, described above as the Graham-Schaefer model. The important point was that the concepts inherent in the Thompson rule were extended to model the relationship between catch and fishing. The idea emerged of an optimum catch, or a maximum sustainable yield, which would sustain fishing as long as need be with no fear of catastrophic decline. The fourth

§ 4.2 THE INTERNATIONAL COUNCIL DURING THE THIRTIES 41

development was a consequence, the idea that the meshes of the trawl cod ends could be enlarged to allow the little fish to escape and grow before they were caught again. The successful demonstration was due to the fishing skipper of the English research vessel *George Bligh* and by the mid-thirties an early form of mesh regulation was in force in the British trawling fleet in the North Sea.

The four developments were all linked in the rather sudden realization that the problem, which had faced fisheries biologists for thirty to forty years, was soluble. Meetings were arranged to bring together the nations exploiting both fishes and whales and to make some international structure analogous to the International Halibut Commission. The arrangements were overtaken by the Second World War, but as a consequence of the Overfishing meeting of the International Council in Berlin in 1939, many of the projects were in fact established when that war was over: the International Commission for North West Atlantic Fisheries (ICNAF), the Permanent Commission, later to become the North East Atlantic Fisheries Commission (NEAFC), and the International Whaling Commission (IWC). Although the three international agencies did not become active until the fifties, their foundations in science were laid in the thirties.

The source of this activity was a visit made by Russell to Raymond Pearl, the American demographer who had introduced the logistic curve to the study of human populations. It was applied by Graham in his study of the plaice population of the southern North Sea. But Graham had also been much influenced by Thompson's work on the Pacific halibut, particularly by the use of Thompson's rule to achieve regulation by small annual changes in fishing effort and thence in stock density and in catch. The model that Graham produced was framed in stock without the vital parameters named by Russell, because they were then difficult to determine. But Graham's second step was to lay the foundations of tagging experiments as a means of estimating fishing mortality, because the rate of increase of the stock was assumed proportional to fishing mortality. During the Second World War Graham was scientific adviser to the 2nd Tactical Air Force where he practised the then new art of operational research. On his return to fisheries research in peacetime, his pupils, Beverton and Holt, applied this method to the dynamics of exploited populations. Their approach was to return to the Russell equation and synthesize a model from those vital parameters.

The reasons for success were twofold. The first was that by 1936 it was clear that the simple regulation of the Pacific halibut and the Baltic plaice had not only halted decline in stock, but had started recovery as had been hoped if Thompson's rule were reliable. The second reason was that the theory was presented in the simple and clear way that carries conviction. Then when means of calculating maximal or optimal yields became available later, men's minds were receptive to them.

4.3 The development of conservation during the fifties and sixties

In a number of countries, minimum landing sizes had been enforced as national regulations for a long time. The first acts of the two North Atlantic Commissions were to establish minimum mesh sizes for different areas and the minimum landing sizes were related to them. Within a decade or so, such rules became well established for all the trawling fleets in the North Atlantic. During the sixties, the pressure of fishing slowly increased and it became gradually clear that mesh regulation by itself was not enough. It is a least conservation because it benefits the smallest species available for capture. For example, in the North Sea, the minimum mesh size is effectively designed for sole and haddock, but not for larger fish such as cod and plaice. Two steps of some importance were taken by ICNAF. The first was an agreement to enforce the rules of the commission internationally, i.e. a fisheries protection vessel could stop a fishing boat of any ICNAF nation at sea and examine its nets; any infraction was laid before the courts in the fisherman's own country. The second step was to add catch quotas to the existing mechanisms of conservation, which meant partly that finer controls could be used and partly that better conservation could be used for some species. At this time, cohort analysis was introduced into the scientists' repertoire: the catch equation was used separately at each age in a cohort and from the catches in numbers at each age, estimates of fishing mortality became available. From such developments the control of demersal fisheries in the North Atlantic became quite successful on the basis of the yield per recruit model and of cohort analysis. (However, in 1977, most stocks in the North Atlantic remain somewhat overexploited.)

Unfortunately, however, there were failures. In 1955, the long-established East Anglian herring fishery collapsed and within twelve years that fishery no longer existed. Other herring fisheries in the North Sea declined and in 1968 the Atlanto-Scandian herring fishery became extinct. Because the quantities landed in the two fisheries were considerable (more than two million tons each year), the effects of failure extended beyond the fishing industry itself into the processing industry. It is likely that this collapse was due to recruitment overfishing, i.e. recruitment declined absolutely under the pressure of fishing. But the failure was not limited to the herring. During the early sixties, the Arcto-Norwegian cod stock came under very heavy pressure and four year classes in succession were reduced to very low levels indeed and the stock suffered from recruitment overfishing (Fig. 4-2); fortunately there are some signs of recovery at the present time. The source of failure lay in an application of the yield per recruit model, which ignored the dependence of recruitment on parent stock. The variability of recruitment masked the true trend of that relationship.

The failure of the Antarctic blue whale stock was due to the same cause.

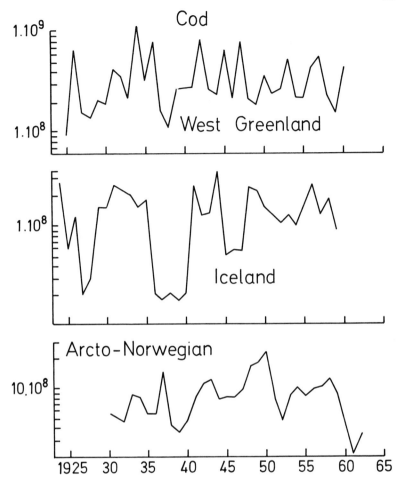

Fig. 4-2 Time series of recruiting year classes in the cod stocks off West Greenland, Iceland and in the Barents Sea (Arcto-Norwegian). (From CUSHING and HARRIS, 1973. With permission from Conseil International pour l'Exploration de la Mer.)

The International Whaling Commission introduced catch quotas, in 1948 and during the fifties, which failed to halt the decline in catches. One of the reasons for this was an argument on age determination. It was no accident that when a committee of fisheries biologists was asked to examine the problem, they should solve it with the Graham-Schaefer model. It took them a little time to convince the commissioners that their clear and simple solution implied a drastic decrease in catch quota. By

1968, when the number of blue whales might have been reduced to about a thousand animals, the commission banned their capture. Whales do not grow much as adults, being mammals, so would not be vulnerable to growth overfishing and as they produce only one calf at a time, must be supremely vulnerable to recruitment overfishing. However, had there been no commission, collapse would have come much earlier and it can be shown that the limited catch quotas, in that sense, saved money.

The Inter-American Tropical Tuna Commission (IATTC) was set up in 1948 and in the eastern tropical Pacific, the yellowfin tuna has been successfully conserved with the use of the Schaefer model. The International North Pacific Commission (INPFC) was established in 1950. The work of INPFC has been concerned with halibut, salmon, king crab and herring; it includes and continues the work of the International Halibut Commission. The main purpose of the INPFC in the past has been to secure the abstention of Japanese fishermen from taking the salmon stock in the Alaska gyral. The abstention principle states that a potential entrant to the fishery should abstain from fishing if the stock is already exploited at the maximum sustainable yield. For example, some herring stocks were shown to be more or less unexploited and the Japanese were free to exploit them if they so desired. One of the great scientific successes of the INPFC was the taxonomic separation of salmon of Asian rivers from those of North American ones, by two distinct parasites and a morphometric analysis; very roughly, Asian salmon lived in the western ocean and North American fish in the eastern one, which confirmed the arbitrary line established much earlier in the history of the Commission. There is one great contrast between the Atlantic and Pacific Commissions; those in the Atlantic aim at the regulation of all fish stocks in that ocean, whereas those in the Pacific serve to conserve only those stocks of interest to North American fishermen.

The main conclusion from this brief history is that the practice of conservation places considerable demands upon the science. No sooner has a model been shown to be successful in one area than a modification is needed in another. The Graham-Schaefer model replaced that of Thompson's rule and itself was replaced with the yield per recruit model. Hardly had this become common practice before a distinction was being drawn between growth overfishing and recruitment overfishing. Today, fisheries biologists concern themselves with cohort analysis to operate the catch quotas and with the problems of recruitment and parent stock.

4.4 Management today

The limit to world catches of fish by conventional means is about one hundred million tons and at the present rate of expansion it might well be reached within the next decade or so. Fleets today are very mobile and they can work from one side of an ocean to the other within a single trip.

§ 4.4   MANAGEMENT TODAY   45

European stern trawlers with freezers aboard work between the Grand Banks and the Barents Sea. Russian trawlers from Kaliningrad base themselves on Cuba to work the South Atlantic and American tuna clippers from San Diego work off Angola. Spaniards exploit the stocks off South Africa and take shrimp off Mozambique. The Japanese and Korean fishermen are seen in all oceans except the high latitudes of the North Atlantic. The combination of a foreseeable limit to catches and the high mobility of expensive and cost-conscious fleets means that management must extend everywhere in the world ocean.

Under the aegis of the Food and Agriculture Organization of the United Nations (FAO), a number of new commissions have been established, for example, off West Africa, off Angola, South Africa and Mozambique and for all the tuna in the Atlantic. Such commissions are worked by national scientists with perhaps some initial assistance from the Fisheries Department of FAO. Another new important commission is the Indian Ocean Fisheries Commission, important because the last major resources are likely to be discovered in that ocean; the sardines off Southern Arabia and off Somalia have already been referred to. The waters between Arabia and the Seychelles are very rich and the area of high production during the southwest monsoon is probably one of the world's largest. In each of the new commission areas, exploitation and conservation may well proceed hand to hand.

However, despite the somewhat restrictive prognosis on a broad scale, exploration continues in such well-known areas as the North Sea and other parts of the North Atlantic. In the last decade, new fisheries for sandeels have been established on the Norfolk banks and in the eastern North Sea. The Norwegians have caught hundreds of thousands of tons of mackerel in the northern North Sea and in the last year or so a Russian fleet of about a hundred vessels has worked for sprats not far from the English coast. There are small cod-like fishes that are exploited for fish meal, principally *Gadus esmarkii*. In the northeast Atlantic the Norwegian fleets have been very active and at the present time are sustained by stocks of capelin on the northern coast of Norway; lightly exploited stocks of such fishes may occur north of Iceland and north of Newfoundland. The most remarkable discovery of recent years is that of a stock of blue whiting that lies in a midwater band (300 to 500 m deep) about 50 to 80 km off the Continental Shelf from the Porcupine Bank off Eire northwards to the Shetlands and it may very well be a very large stock indeed. Yet, however extraordinary it may be that substantial quantities are still discovered in such well explored regions, it remains true that they are found by exploiting the byways of the marine ecosystem.

The oldest International Commission is not a fisheries commission but the Fur Seal Commission and as the oldest it is probably the most successful. Fur seals breed on the Pribilov Islands and on Commander Island in the Bering Sea. The old males come ashore first in spring and stake out the territories which they will defend in the bloodiest way in order to protect their harems. When the females reach the beaches they give birth to the pups conceived a year before. Inland from the beaches lie the grassy areas where the young bachelors play and grow and it was these animals that were killed in large numbers for their skins. During the last century catches reached very low levels, but since 1911 when the Commission was established catches have gradually increased. The reason for this is that the owners of the islands (the Pribilov Islands belong to the United States and Commander Island to the Soviet Union) are able to control catches and estimate stock on the beaches with considerable accuracy. The stock was allowed to build up slowly until the pups began to suffer somewhat higher mortality indicating that some form of density dependent control mechanism was at work. From a form of stock/recruitment relationship, a maximum yield was calculated and this remains the present object of the Commission's work. The reasons for the success of this model Commission is first that the stock is well and clearly defined, secondly, that catch and stock in age can be precisely separated and, thirdly, the control in catch was very easy to enforce.

The lesson for the fisheries commissions is obvious—the three conditions for the success of the Fur Seal Commission should be observed. However, it is sometimes hard to establish stock unity in fishes, although a start has been made in that direction. But given a good statistical basis on the market the estimation of stock and catch is becoming easier. Lastly, the measures of regulation need enforcement, as was also shown in the North Atlantic. The use of the international fisheries commissions has, however, sometimes been questioned, partly for particular political reasons, but mainly because the slow historical trend towards good conservation has not been detected in the day-to-day arguments. There is a simple test of their value. In the last forty years of the last century, many Royal Commissions investigated the conditions of the industry in the United Kingdom, particularly the declining stocks. Since the Second World War no such investigation has been needed. If there have been failures like that of the northeast Atlantic herring, there has been considerable success in conservation in the North Atlantic.

5 Science and the Fisheries Today

5.1 The structure of currents in the sea

As described in an earlier chapter, there is a permanent structure of currents in the world ocean. In the shallow sea, however, like the North Sea and English Channel, the existence of such currents is overlain by the tides. In the very shallow waters between the Dogger Bank and the Isle of Wight, the tides are very swift, the sea bed coarse and the water very turbid. To the north of the Dogger or west of the Isle of Wight, the water is deeper and clearer, the sand is fine and the tides are slower. In this deeper water a thermocline forms in summer, but never in the shallower region where tides are swift. However, the tides provide a direction and a velocity which is as permanent as the main current structures of the world ocean.

Fishes exploit the structures of tide and current to migrate across considerable distances in the sea. With acoustic tags that report their presence and that of the fish to which they are attached to a highly sophisticated sonar with a very narrow beam, it is possible to follow single fishes for a number of days. In the region of swift tides in the southern North Sea, plaice lie on the bottom during northbound tides and 'fly' in midwater on southerly tides as the populations move southward to spawn (Fig. 5-1). After the spawning season, the behaviour is reversed. It is not very clear how the physiological mechanisms work, but the plaice are obviously able to migrate across considerable distances with the least expenditure of energy. On hydrodynamic grounds fish should swim at about one length/second; to cross an ocean quickly, in the way that tuna do, fish must find a window in the velocity distributions of ocean currents. Albacores have crossed the Pacific in as little as sixty days and at one length/second such a journey is impossible, so they may find ways of using the faster streams to their advantage.

It is possible that our picture of the distribution of currents in the sea is somewhat over-simple; for example, the Gulf Stream is not merely a river of swift current flowing directly north from the Straits of Florida towards Cape Hatteras and points east, but is one of meanders shedding great eddies as it spreads into the North Atlantic Drift. An extension of this current flows between the Faroe Islands and the Shetland Islands off the coast of Norway to Spitzbergen and eastwards of North Cape. Off the Norwegian coast, the current splits into a set of elongated swirls, parts of which are northbound and parts southbound, a much more complicated structure than indicated by a simple arrow on the chart. Another indication of complexity is shown by the traces of caesium 137 (at very low levels indeed) from the nuclear power station at Windscale in the Irish

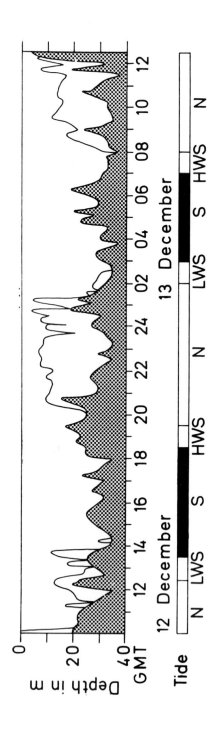

Fig. 5–1 The vertical movement of a single plaice, during a tidal cycle, shown by an acoustic tag which responds to the Admiralty Research Laboratory scanner. The hatched area represents the sea bed in profile and the movements of the tagged plaice are shown as full lines; associated are the tidal streams, the southerly streams being shown in black. (With the permission of F. R. Harden Jones.)

Sea. The caesium water passes from the Irish Sea into the Atlantic west of Scotland and flows northward in a narrow coastal band round the north coast of Scotland into the North Sea; off the Firth of Forth, it swings eastward into the open sea. The important point is that the caesium stream retains its identity for a considerable distance apparently defying the diffusive processes in the sea.

Physical oceanographers have had to describe motion in the sea on a broad scale in order to obtain understanding. The phenomena indicated above are on a lesser scale but reveal unexpected complexities. Although fish can make transoceanic migrations, their usual range lies within that of an oceanic structure such as the Norwegian Sea, the North Sea or the Alaska gyral. Haddock may ride the 'caesium stream' from the west of Scotland into the North Sea and it would certainly provide them with a convenient and lazy passage. Such local and undiscovered structures could be well exploited for short periods by fish on passage.

In the future, we might expect that the detailed studies of migration with sonar and acoustic tags will be combined with equally detailed studies of water flow which would lead to considerable understanding of the migration of fishes. The particular point that might well emerge is the degree to which fish use tide or current to make a migration and the degree to which they ignore them. Fishermen have long known that fisheries occur at particular places at certain times of the year with persistent regularity. The link between studies of migration and those of water flow will elucidate such regularities and provide a basis for understanding.

5.2 Catch statistics and population dynamics

Much has been achieved with the present levels of catch statistics which are usually collected on a voluntary basis. The British system is based on the total catch for a voyage from one ground together with the number of hauls made, and from such information a good estimate of average catch or catch per effort can be obtained. This system, introduced more than half a century ago, is no longer as useful as it was, because on a single voyage fishermen may change their grounds on a number of occasions. The ideal arrangement is to collect individual catches by position of capture as the Japanese do from their world-wide oceanic tuna fleet. When grouped by 5° squares the information is of high quality which is perhaps why some of the recent Japanese work on the analysis of such statistics is so good.

Some idea of the general quality of the material available to fisheries biologists is given in Figure 5-2. We have information on catch per effort (or stock density) of cod for many regions in the North Atlantic. A model was constructed which simulated the changes in stock density from year to year in terms of the vital parameters of the population. Some of the

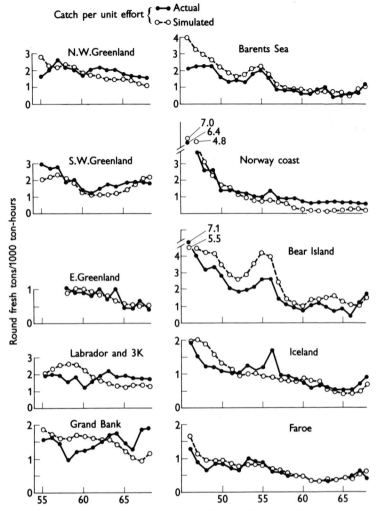

Fig. 5-2 A comparison between observed (●) stock densities of cod in the North Atlantic and simulated ones (O) in various areas. (CLAYDEN, 1972.) (British Crown Copyright. Reproduced with the permission of HMSO.)

comparisons between model and stock density are poor, such as the Grand Bank, but others are good (Barents Sea and Faroe Islands). We conclude that the statistics on which the models are based are quite good, even if improvements are often needed.

In temperate waters the age structures of fish populations are well described and consequently yield per recruit models may be constructed quite readily; in tropical seas, fish are sometimes hard to age and so growth rates and death rates cannot be estimated quite as readily. Cohort analysis (or virtual population analysis) yields one matrix by ages and years of fishing mortality and another of stock, both of which are accurate up to three years before the last age in the cohort. Hence, there is some difficulty in forecasting a catch quota ahead of well-established data. The ideal way out of this impasse would be to set a quota which remained stable for a number of years. However, nearly all stocks of fish in the North Atlantic are overexploited and the prime need is to reduce fishing activity considerably, by perhaps a third or a half, but which of course can only be cut by small fractions. The scientific problem may be to use the quota as a source of information in the system of analysis.

However, the major scientific problem facing fisheries biologists is the dependence of recruitment on parent stock. At the present stage of development a decade or two of observations at high, middle and low stock may generate a relationship which is credible but not always convincing, but to improve it many more decades of observations are needed. The real danger is that some yield per recruit curves, particularly the flat-topped ones, may conceal the point at which recruitment becomes reduced by heavy fishing. A number of devices are used to overcome this failure but the real need is to understand the processes by which the annually recruiting year classes are generated.

The Ricker curve, as used today, estimates density dependent and density independent mortality between hatching and recruitment as an average from the observations of recruitment and parent stock. In an ideal world such parameters would be estimated independently and the curve would enjoy the practical success of the yield per recruit model. Such a model would not depend on the accretion of observations decade by decade but on the errors of estimation of the independently determined parameters.

To put the problem as simply as this is to underestimate the difficulty of establishing such parameters independently. What is really needed is a considerable body of knowledge on the growth and death of fish larvae and young fish of many species. To acquire such knowledge is both expensive and laborious, for many ships would have to work long periods. The material on which the apparently simple yield per recruit model was based was collected with considerable patience during the twenties and thirties. We are slowly proceeding through the same period of collection as we approach the solution to the problem of recruitment and parent stock.

5.3 Independent estimates of stocks by acoustic methods

When fishermen first bought echosounders they did so to record the depths of the sea. Such machines make quasi-continuous charts of the sea bed which resemble sections of the sea from the surface to the bottom.

Shoals of fish on the sea bed or in midwater are shown quite clearly and the fishermen were very quick to take advantage of this observation. If such records are regarded as sections of the sea, it is quite easy to survey the fish in the sea with an echosounder. Such a survey is physically valid if the acoustic output is constant, if the water is not too deep, if fish live at about the same depth and if the sea is not too rough. During the early fifties many such surveys were made.

They estimated relative abundance in area and were of considerable exploratory value, but it would obviously be desirable to estimate absolute abundance. The first step was to recognize single fishes; the echosounder transmits a pulse of sound of about one millisecond in duration (or about 0.75 m in range) and any signal received for some transmissions must have been recorded from a discrete target or a single fish. With a suitable echosounder a large cod might be detected quite decisively in depths as great as 500 m. The Norwegians were able to survey the Barents Sea twice a year for two years and describe the distribution of single cod in some detail.

The next step forward was to establish the volume in which the single fish lived which is necessary to establish the absolute abundance of the population. The total volume sampled by an echosounder is pear-shaped with the base of the pear at maximum range. This limit in range is determined by the noise in the water due to the ship's propeller, the roughness of the sea and the transmission of the echosounder itself. The beam of an echosounder is superficially like a headlight, brightest at the centre and dimmer at the edges. So if a large cod is detected at 500 m at the centre of the beam it must be detected at a lesser range at the edge. The pear shaped volume is determined by the product of the maximum range and the beam pattern.

Sound is scattered or reflected from different parts of the fish; flesh, scales, bones and particularly from the swim bladder. Indeed half the signal comes from the swim bladder which occupies only one-twentieth of the fish's volume. Hence a survey of fishes such as cod with swim bladders would be probably more successful than one of bladderless fishes, like mackerel. But much more important, the signal from a single fish is proportional to its weight (or somewhat less than it). Hence arises the possibility of distinguishing fishes by size, but the variability of the experimental results is high. Fisheries biologists would like to estimate fish populations by sizes in absolute abundance independently of commercial catches, which can be done by acoustic survey. Such a survey of hake off South Africa and South West Africa has shown that numbers of fish can be estimated (Fig. 5–3) and at the time it was possible to make a roughly estimated size distribution.

Fishes cannot be identified acoustically and any echo survey must be supported by capture with trawl hauls. The great advantage of such a survey is that the sampling power of a trawl survey is increased

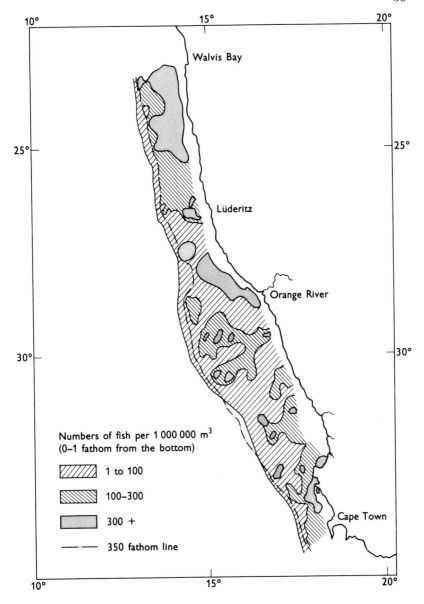

Fig. 5-3 The distribution of hake off South Africa as shown by echo survey within one fathom of the sea bed. (CUSHING, 1967.) (British Crown Copyright. Reproduced with the permission of HMSO.)

enormously with the use of acoustic information. The method is of considerable use in the exploration of unexploited stocks and has been so used extensively. But it has not yet developed far enough to provide the independent stock estimate that the fisheries biologists need so badly. There are two statistical reasons for this: the first is that the differences in sizes of fish from area to area cannot yet be taken into rigorous account, and the second is that the integrator which records biomass may be open to signals from the biomass of abundant very small and unwanted animals.

5.4 Long term trends in the climate

On a time scale of centuries distinct changes in the climate have occurred in different parts of the world. For example, in the eleventh and twelfth centuries there was a climatic optimum, a warm period when the weather was persistently better than at any other time since the birth of Christ. The seventeenth and eighteenth centuries were cool, so cool that the period has been called the Little Ice Age when the Thames frequently froze in winter and when the summers must have been uncomfortably short. Between the climatic optimum and the Little Ice Age the weather was variable, but never very good. In the early nineteenth century the climate improved considerably, but between 1870 and about 1920 there was a period of climatic deterioration. In the twenties, thirties and forties of the present century the climate improved again in northwest Europe when the summers were long and the winds westerly and when the waters of the northeast Atlantic became warmer by as much as 1°C, an event of some significance. Subsequently between 1945 and 1971 the climate deteriorated again. Such are the scales of climatic change in western Europe.

We are very familiar with the depressions and anticyclones that track across the British Isles usually from west to east. There is a rough pattern in the planetary distribution of both, there being two to six major atmospheric features in the northern hemisphere at any one time. Tropical air is warmer than polar air and it expands more, so a pressure gradient develops between high and low latitudes. The westerlies blow across this gradient on average at a middle latitude. As the earth spins the air is moved in swirls and eddies. A cold trough is generated by a movement towards the equator and a warm ridge by a poleward one. In this way the pattern of anticyclones and depressions is generated, all of which move on average from west to east. The weather is often considered to vary quite irregularly and indeed perhaps randomly. But there are regularities in the sequence of meteorological events which we term climate; for example, in recent years northerly winds have blown over the British Isles in April, persistently from year to year.

Recently it has been shown that the great mass of water in the

§ 5.4   LONG TERM TRENDS IN THE CLIMATE   55

subtropical northern Pacific tends to stabilize climatic events. The long hot summer of 1959 in the British Isles occurred some little time after a region in the North Pacific had retained heat for an abnormally long period. This region is the largest area in which heat can be conserved and it acts as a climatic stabilizer; the '1959 event' persisted for eighteen months to two years. As the air streams eastward from the North Pacific it crosses the Rockies and eastern Canada. The first tends to form a persistent anticyclone and the second a persistent depression, for recurring periods. In the North Atlantic the winds blow between the Azores High and the Iceland Low. Such winds are westerly.

However, there is a periodicity of about 100–110 years in such systems. The winds blow across a pressure difference between Iceland and the Azores; if high the wind is strong and vice versa. As the pressure difference reduces, the wind slackens and then in western Europe it may be blocked by an anticyclone and then southerly winds predominate. Figure 5–4 shows how the winds have shifted from west to south and back again in the last century or so.

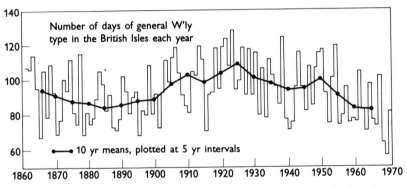

Fig. 5–4   The changes in westerly wind direction over the British Isles for a long period. (From LAMB, H. H. *Climate, Past, Present and Future Vol. 1*, 266. With permission from Associated Book Publishers Ltd, London.)

Thus there is some reason to associate such events with analogous periods in the fisheries. The alternation between the Norwegian herring fishery and that off the Bohuslån coast of Sweden since the Middle Ages indicates such a periodicity. There is a similar periodical fluctuation in the fishery for the Japanese sardine which suggests that events might be connected throughout the northen hemisphere, as might be expected if climatic changes are really driven by the heat sink in the northern subtropical Pacific.

The secular trends in wind strength and direction are very significant from the point of view of the variation from year to year in recruitment. The match or mismatch of larval production to that of their food might well be governed by the differences in wind strength and direction from year to year. This cannot yet be shown because it is a little difficult to extract the right information for spawning grounds defined in rather precise geographical terms. However, year class strength of some fishes has been shown to be correlated with differences in temperature at fixed positions (Fig. 5–5).

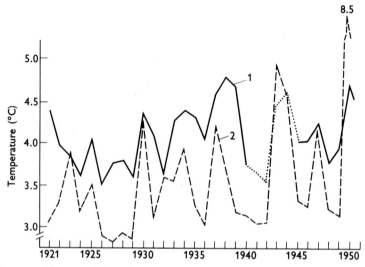

Fig. 5–5 The correlation of year classes in the Norwegian herring stock (2) with differences in temperature (1) on the Kola meridian (in the Barents Sea). (The dotted line represents temperatures interpolated from other information during the Second World War.) (From ISHEVSKII, G. K. (1964). *The systematic basis for predicting oceanological conditions and the reproduction of the fisheries.* Moscow, 165 pp.)

It is very unlikely that recruitment is affected by salinity directly, but the differences in saltness may reflect changes in water movement, which are themselves associated with changes in the wind. However, this must remain a speculation; much more important is the possibility of forecasting year class strength from simple observations that are collected routinely from 'bateaux routiers' as they cross the North Sea. It would be reasonable to suppose that such detailed relationships were explicable on the same basis as the long term alternation of fisheries across the centuries. The existence of a fishery depends upon the generation of good year classes persistently for many years.

5.5 Fish farming

Carp were grown in monastery stews throughout the Middle Ages and the ponds were rotated yearly in three stages of growth; fry and fingerlings, yearlings, and grown fish. A fourth pond was left fallow for a year. Similar culture practices for carp and other fresh water fish have been developed in Europe and Israel and for rainbow trout in the United States. In Indonesia and other Far Eastern countries there are traditional intense methods of production in rice fields or in fertilized ponds.

In the sea, fish farming is mainly in an experimental or developmental stage. In the Inland Sea of Japan, however, prawns are reared in quite large quantities and yellowtails are grown in fenced areas on the natural plankton in the sea. Perhaps the most important development in marine fish farming is the culture of salmon whereby the return of adults to the parent stream is exploited. The method was developed at the University of Washington in Seattle; sockeye smolts were released from the hatchery there and adults returned to it two or three years later. The growth is put on at sea with natural food and there is no need to buy protein food to convert to fish protein, which is the major disadvantage of fish farming. However, the price of salmon is high enough to exploit such conversion and in the future, expensive fish such as turbot might be reared in sea water on cheap protein food.

5.6 Pollution

As certain industries have developed since the Second World War, pollutants have appeared in the air, in rivers, in estuarine sediments and in the sea. They are mainly radioactive wastes from the nuclear industries, organochlorines from the plastics industries and heavy metals. They are distributed pervasively in the atmosphere, in the ocean and in rivers to such a degree that it is rather hard to find control environments where they cannot be detected.

The most important problem is to establish the levels of pollutants in the food fishes and then to agree international limits to such quantities. As background to such values the quantities of all pollutants are measured regularly in 'baseline studies' in various parts of the world. Research continues into the concentration of different pollutants into varied parts of the food chains. The main result from all this work has been to show that although pollutants are to be found everywhere in the sea, the food fishes are safe to eat and it is probable that most vital processes in populations and ecosystems have not been damaged.

5.7 Conclusion

The choice of subjects in this chapter is inevitably a personal one of mine. However, they do represent together something that resembles the present tide of advance of fisheries science. Some of it is strictly practical like the work on pollution, but much of it is really basic research on the study of wild populations. No feral populations have been studied in such detail for such long periods of time and this information is starting to yield results of considerable importance.

There is, however, a deeper motive for the practice of fisheries research. There are many forms of practical ecology ranging from wildlife management, such as the conservation of fish stocks, to the preservation of endangered species. The first has a purely economic end in view and the second has primarily a moral objective, that it is wrong to extinguish a species such as the Arabian oryx. Fisheries conservation is a successful practice with a strictly economic end based on the science described here and it is being employed throughout the world ocean.

At the present time many of the stocks of fish in the North Atlantic are overexploited. This means that the same or greater catch can be obtained with perhaps half to two-thirds of the present number of fishermen. The Commissions are organizations to which nations adhere quite voluntarily; in other words they wish to agree to the courses of action proposed. It follows that the Commissions proceed rather slowly, rather more slowly than many people would like. However, progress is made, even if some fisheries have become extinguished because the Commissions' progress had been too slow.

As a consequence the practice of a fisheries biologist is very varied. In one week he may work in one of the International Commissions providing advice on the details of catch quotas which are being 'horse traded' between the national representatives. In another he might take part in an international working group on the state of a particular cod stock; the data from different countries are collated, the population statistics are analysed and the group eventually agrees on a quota, based on a predetermined management objective, e.g. at the maximum sustainable yield. At some other time of the year, the fisheries biologist may find himself at sea in a small or large research vessel; Figure 5–3 is the chart of an acoustic survey which I did off South Africa in H.M.S. *Hecla*, a naval survey vessel. Such cruises may last for as short a period as a week or ten days or for as long as five or six weeks. The ships range around the waters of the British Isles, off the beaches, off the continental shelves, in the distant waters of the Barents Sea or even on some occasions as far away as the Antarctic.

The work is varied and demanding and is intended to secure the livelihoods of fishermen. Some parts of the work appear to be remote from the deck of a trawler but the records of catches are the bones of any population study. To be successful the fishermen must store considerable knowledge of the natural history of the fishes they catch and such information is of great value to the biologists. Hence they work with the fishermen and learn from them to maintain the stocks of fish in the sea at a profitable and sustainable level.

References

VON ARX, W. (1962). *An Introduction to Physical Oceanography*. Addison-Wesley Reading, Mass., U.S.A.

BARANOV, F. I. (1918). On the question of the biological basis of fisheries. *Nauchnyi. issledovatelskii iktiologicheskii Institut, Izvestiia*, 1, 81–128.

BEVERTON, R. J. H. and HOLT, S. J. (1957). On the dynamics of exploited fish populations. *Fishery Invest., Lond.*, Ser. 2, 19.

CLAYDEN, A. D. (1972). Simulation of the changes in abundance of the cod (*Gadus morhua* L.) and the distribution of fishing in the North Atlantic. *Fish. Invest. London Ser.* 2, 27(1), 58 pp.

CUSHING, D. M. (1967). The abundance of hake off South Africa. *Fish. Invest. London Ser.* 2, 25.10, 20 pp.

CUSHING, D. H. (1975). *Marine Ecology and Fisheries*. Cambridge University Press, London.

CUSHING, D. H. and HARRIS, J. G. K. (1973). Stock recruitment and the problem of density dependence. *Rapp. Proces-Verb. Cons. int. Explor. Mer.*, 164, 142–55.

FUKUDA, Y. (1962). On the stocks of halibut and their fisheries in the northeast Pacific. *Intern. N. Pacific Fish Commn. Bull.*, 7, 39–50.

GARSTANG, W. (1900–3). The impoverishment of the sea. *J. mar. biol. Ass. U.K.*, NS 6, 1–69.

GRAHAM, M. (1935). Modern theory of exploiting a fishery, and application to North Sea trawling. *J. Cons. int. Explor. Mer.*, 10, 264–74.

GRAHAM, M. (1938). Rates of fishing and natural mortality from the data of marking experiments. *J. Cons. int. Explor. Mer.*, 13, 76–90.

GULLAND, J. A. (1971). *The Fish Resources of the Ocean*. Fishing News (Books) Ltd, West Byfleet, Surrey, England.

HEINCKE, F. (1913). Untersuchungen über die Scholle Generalbericht. I. Schollenfischerei und Schonmassregeln. Vorlaufige Kurze Ubersicht über die Wichtigsten Ergebnisse des Berichts. *Rapp. P.-v. Réun. Cons. perm. int. Explor. Mer.*, 16, 1–70.

HJORT, J., JAHN, G. and OTTESTAD, P. (1933). The Optimum catch. *Hvalråd. Skr.*, 7, 92–127.

MERRIMAN, D. (1941). Studies on the striped bass (*Roccus saxatilis*) of the Atlantic Coast. *Fishery Bull. Fish Wildl. Serv. U.S.*, 50, 1–77.

PARSONS, T. R. and TAKAHISHI, M. (1973). *Biological Oceanographical processes*. Pergamon Press, London.

PETERSEN, C. G. J. (1894). In the biology of our flat-fishes and on the decrease of our flat-fish fisheries. *Rep. Dan. biol. Stn.*, 4.

RICKER, W. E. (1958). Handbook of computations for biological statistics of fish populations. *Bull. Fish. Res. Bd Can.*, 119.

RUSSELL, E. S. (1931). Some theoretical considerations on the 'overfishing' problem. *J. Cons. int. Explor. Mer.*, 6, 3–20.